BACTERIAL ARTIFICIAL CHROMOSOMES

Edited by **Pradeep Chatterjee**

INTECHOPEN.COM

Bacterial Artificial Chromosomes
Edited by Pradeep Chatterjee

Published by InTech
Janeza Trdine 9, 51000 Rijeka, Croatia

Edition 2014

Publishing Process Manager Daria Nahtigal

Technical Editor Teodora Smiljanic

Cover Designer InTech Design Team

Image Copyright Vphoto, 2011. Used under license from Shutterstock.com

Additional hard copies can be obtained from orders@intechopen.com

Bacterial Artificial Chromosomes, Edited by Pradeep Chatterjee
p. cm.
ISBN 978-953-307-725-3

INTECH

open science | open minds

Contents

Preface

It has been a little over two decades since the stable propagation of 100 kb-sized DNA in bacteria by Drs. Nancy Shepherd and Nat Sternberg using the phage P1 packaging system. The Bacterial Artificial Chromosome (BAC) system was developed soon after by Drs. Hiroaki Shizuya, Bruce Birren, Ung-Jin Kim, Melvin Simon and colleagues. Genomic DNA libraries are easier to construct using electroporation, instead of P1 packaging, and clones can propagate DNA of much larger size using the BAC system. As a consequence, BACs became very popular among researchers in the genome community and Drs. Pieter de Jong, Kazutoyo Osoegawa, Chris Amemiya and their colleagues generated a series of genomic DNA libraries from several vertebrate organisms that are not only of much higher coverage of their respective genomes but also comprised of clones that had DNA inserts of larger average size. These libraries played important roles in the assembly of genome sequences of several vertebrate organisms including the human, mapping genes and genetic markers on chromosomes, and serving as useful tools in comparative genomics studies of related species. A chapter representative of such applications of BAC libraries is included in this book.

The past decade witnessed the wide spread use of clones from BAC libraries of numerous organisms for functional studies. The large insert DNA size and easy maneuverability of that DNA in bacteria has contributed to the growing popularity of BACs in transgenic animal studies. The realization that many control elements of genes important during vertebrate development are actually located at large distances along the DNA from the coding sequences of the gene have made BACs increasingly indispensable for studies of developmentally regulated genes using transgenic animals. A different area of interest arose from the same attractive features of BACs, and relates to their use as vectors for cloning the very large genomes of several DNA viruses. Faithful propagation and easy mutational analyses of the BAC-viral DNA in bacteria allowed rapid assignment of function(s) to the numerous open reading frames in the viral genome when that BAC-viral DNA was reintroduced into permissive hosts for a productive infection. Several chapters of this book illustrate the variety of applications in this area.

Several new technologies have been developed to alter sequences in BAC DNA within its bacterial host. While all of these methods utilize DNA recombination of some sort, the more widely used ones require re-introducing homologous recombination function of E.coli or phage λ back into the severely recombination deficient host. This book also contains a couple of chapters illustrating the usefulness of BACs in functionally mapping gene regulatory elements. In this context the recent demonstration by Dr. Koichi Kawakami and colleagues that the vertebrate transposon system Tol2 can be re-engineered to facilitate integration of BAC DNA into the chromosomes of zebrafish and mice is likely to accelerate the use of BACs in a variety of studies with transgenic animals.

This book focuses on the numerous applications of Bacterial Artificial Chromosomes (BACs) in a variety of studies. The topics reviewed range from using BAC libraries as resources for marsupial and monotreme gene mapping and comparative genomic studies, to using BACs as vehicles for maintaining the large infectious DNA genomes of viruses. The large size of the insert DNA in BACs and the ease of engineering mutations in that DNA within the bacterial host, allowed manipulating the BAC-viral DNA of Varicella-Zoster Virus. Other reviews include the maintenance and suitable expression of foreign genes from a Baculovirus genome, including protein complexes, from the BAC-viral DNA and generating vaccines from BAC-viral DNA genomes of Marek's disease virus. Production of multi-purpose BAC clones in the novel Bacillus subtilis host is described, along with chapters that illustrate the use of BAC transgenic animals to address important issues of gene regulation in vertebrates, such as functionally identifying novel cis-acting distal gene regulatory sequences.

Pradeep K. Chatterjee
Associate Professor
Biomedical/Biotechnology Research Institute
North Carolina Central University, Durham
USA

1

BAC Libraries: Precious Resources for Marsupial and Monotreme Comparative Genomics

Janine E. Deakin
The Australian National University
Australia

1. Introduction

Over the past decade, the construction of Bacterial Artificial Chromosome (BAC) libraries has revolutionized gene mapping in marsupials and monotremes, and has been invaluable for genome sequencing, either for sequencing target regions or as part of whole genome sequencing projects, making it possible to include representatives from these two major groups of mammals in comparative genomics studies. Marsupials and monotremes bridge the gap in vertebrate phylogeny between reptile-mammal divergence 310 million years ago and the radiation of eutherian (placental) mammals 105 million years ago (Fig. 1). The inclusion of these interesting species in such studies has provided great insight and often surprising findings regarding gene and genome evolution. In this chapter, I will review the important role BACs have played in marsupial and monotreme comparative genomics studies.

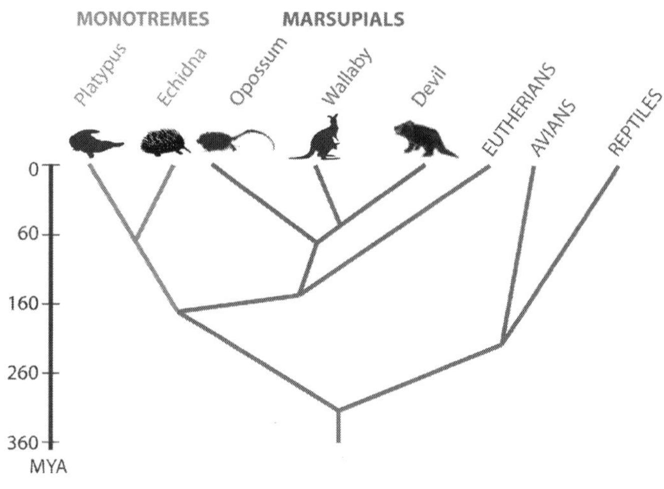

Fig. 1. Amniote phylogeny showing the relationship between 'model' monotreme and marsupial species used in comparative genomic studies.

1.1 Monotreme BAC libraries

Monotremes are the most basal lineage of mammals (Fig. 1), diverging from therian mammals (marsupials and eutherians) around 166 million years ago (mya) (Bininda-Emonds et al., 2007). Like all other mammals, they suckle their young and possess fur, but their oviparous mode of reproduction and their rather unique sex chromosome system are two features of most interest to comparative genomicists. BAC libraries have been made for two of the five extant species of monotremes, the platypus (*Ornithorhynchus anatinus*) and the short-beaked echidna (*Tachyglossus aculeatus*). These species last shared a common ancestor approximately 70 mya. The platypus genome, consisting of 21 pairs of autosomes and 10 pairs of sex chromosomes, has been sequenced (Warren et al., 2008) and a male and a female BAC library constructed (see Table 1). Similarly, the echidna genome has nine sex chromosomes and 27 pairs of autosomes, with a male BAC library available for this species (Table 1).

Species	Library Name	Sex	Average insert size (kb)	Number of Clones
Platypus	CHORI_236	Female	147	327,485
Platypus	Oa_Bb	Male	143	230,400
Short-beaked echidna	Ta_Ba	Male	145	210,048

Table 1. Available monotreme BAC libraries

1.2 Marsupial BAC libraries

Marsupials, a diverse group of mammals with over 300 extant species found in the Americas and Australasia, diverged from eutherian mammals approximately 147 mya (Bininda-Emonds et al., 2007) (Fig. 1). They are renowned for their mode of reproduction, giving birth to altricial young that usually develop in a pouch. Three species of marsupials were chosen as 'model' species for genetics and genomics studies 20 years ago: the grey short-tailed South American opossum (*Monodelphis domestica*) representing the Family Didelphidae, the tammar wallaby (*Macropus eugenii*) from the kangaroo family Macropodidae and the fat-tailed dunnart (*Sminthopsis macroura*) as a member of the speciose Family Dasyuridae (Hope & Cooper, 1990). The opossum, the first marsupial to have its genome sequenced (Mikkelsen et al., 2007), is considered a laboratory marsupial and has been used as a biomedical model for studying healing of spinal cord injuries and ultraviolet (UV) radiation induced melanoma (Samollow, 2006). The tammar wallaby has also recently had its genome sequence (Renfree et al., 2011) and has been extensively used for research into genetics, reproduction and physiology. Although there have been a few studies carried out on the fat-tailed dunnart, the recent emergence of the fatal devil facial tumour disease (DFTD) has led to the Tasmanian devil replacing it as the model dasyurid, with many resources being made available, including genome (Miller et al., 2011) and transcriptome sequence (Murchison et al., 2010). These model species represent three distantly related marsupial orders, with comparisons between these species being valuable for discerning the features that are shared among marsupials and those that are specific to certain lineages. BAC libraries have been made for all four species mentioned above and are summarized in Table 2. The three current model species will herein be referred to simply as opossum, wallaby and devil.

In addition to the model species, BAC libraries have also been constructed for the Virginia opossum (*Didelphis virginiana*), another member of the Family Didelphidae and the Northern brown bandicoot (*Isoodon macrourus*) (Table 2) from the Family Peramelidae. The phylogenetic position of the bandicoots, located at the base of the Australian marsupial radiation, and some of their more unique features make them interesting animals to study (Deakin, 2010). They possess the most invasive placentas among marsupials, with an allantoic placenta more like that found in eutherians, which would make them a valuable species in which to study genomic imprinting. They also deal with dosage compensation in an unusual way by eliminating one sex chromosome in somatic cells (Hayman & Martin, 1965; Johnston et al., 2002).

Species	Library Name	Sex	Average insert size (kb)	Number of Clones
Didelphis virginiana	LBNL-3	Female	170	148,162
Isoodon macrourus	IM	Male	125	
Macropus eugenii	ME_KBa	Male	166	239,616
Macropus eugenii	Me_VIA	Male	108	55,000
Monodelphis domestica	VMRC-6	Male	155	276,480
Monodelphis domestica	VMRC-18	Female	175	364,800
Sarcophilus harissii	VMRC-49	Male	140	258,048
Sarcophilus harissii	VMRC_50	Male	140	165,888
Sminthopsis macroura	RZPD688	Male	60	110,592

Table 2. Marsupial BAC libraries

2. BACs used for gene mapping and sequencing of target regions

Prior to the availability of BAC libraries for marsupials and monotremes, gene mapping by fluorescence in situ hybridization (FISH) was an arduous task, which relied on the isolation of the gene of interest from a lambda phage genomic library. The construction of BAC libraries for the species listed above has facilitated the mapping of many marsupial and monotreme genes by FISH. Initially, PCR products were used to screen these BAC libraries for genes of interest but more recently overgo probes (overlapping oligonucleotides) have proven to be the method of choice, permitting the isolation of many genes from one screening, thereby facilitating the rapid construction of gene maps. Likewise, before the availability of genome sequence, isolating and sequencing BACs containing genes of interest proved a very useful method for obtaining sequence from particular regions of interest. In some cases, even after whole genome sequencing had been performed, it proved necessary to take this targeted approach. These mapped or sequenced BACs have led to a number of important findings, with examples of those having had a significant impact on previously held theories reviewed here. Examples include the determination of the origins of monotreme and marsupial sex chromosomes, the evolution of regions imprinted in eutherian mammals, the unique arrangement of the Major Histocompatibility Complex (MHC) in the tammar wallaby and the evolution of the α- and β-globin gene clusters. BACs have played a vital role in many more studies using gene mapping and/or target region sequencing than can be included in detail in this review and hence, other studies that have utilized BACs for these purposes are listed in Table 3. This is not an exhaustive list but an indication of the breadth of studies in which BACs have played a role.

Species	Genes or Region	Purpose	Reference
Echidna and Platypus	*SOX 3*	Mapping	(Wallis et al., 2007b)
Opossum (*M.domestica*)	Immunoglobulins	Mapping	(Deakin et al., 2006a)
Opossum (*M.domestica*)	T cell receptors	Mapping	(Deakin et al., 2006b)
Platypus	*DMRT* cluster	Sequencing	(El-Mogharbel et al., 2007)
Platypus	Defensins	Mapping	(Whittington et al., 2008)
Platypus	*SOX9* and *SOX10*	Mapping	(Wallis et al., 2007a)
Platypus	Sex determination pathway genes	Mapping	(Grafodatskaya et al., 2007)
Dunnart	*LYL1*	Sequencing	(Chapman et al., 2003)
Tammar wallaby	Prion protein gene	Sequencing	(Premzl et al., 2005)
Tammar wallaby	Immunologulins & T cell receptors	Mapping	(Sanderson et al., 2009)
Tammar wallaby	Mucins & Lysozyme	Mapping	(Edwards et al., 2011)
Tammar wallaby	*SLC16A2*	Sequencing	(Koina et al., 2005)
Tammar wallaby	*BRCA1*	Mapping	(Wakefield & Alsop, 2006)
Tammar wallaby	Cone visual pigments	Sequencing	(Wakefield et al., 2008)

Table 3. Studies in marsupial and monotreme comparative genomics that relied on BAC clones.

2.1 Origins of marsupial and monotreme sex chromosomes

Determining the evolutionary origins of marsupial and monotreme sex chromosomes was the driving force behind much of the gene mapping conducted in these species. The earliest gene mapping work showed that at least some genes found on the human X chromosome were also on the X in marsupials, resulting in the hypothesis that the X chromosome of these two mammalian groups had a common origin. Gene mapping using heterologous probes and radioactive in situ hybridization (RISH) supported the extension of this hypothesis to include monotremes. However, it was only when BAC clones became available for gene mapping that the true evolutionary history of sex chromosomes in these species was revealed.

2.1.1 The marsupial X chromosome

Like humans, marsupial females have two X chromosomes whereas their male counterparts have a X and a small Y chromosome, meaning that they require a mechanism to compensate for the difference in dosage of X-borne genes between females and males. Several decades ago, it was shown that several X-linked genes in human were also located on the X in marsupials and one X chromosome was inactivated in somatic cells to achieve dosage compensation. However, even in these early studies, striking differences in the characteristics of X inactivation in eutherians and marsupials were evident. Marsupials were found to preferentially silence the paternally derived X chromosome rather than subscribing to the random X inactivation mechanism characteristic of eutherian mammals. This inactivation was found to be incomplete, with some expression observed in some tissues from the inactive X and thus, appeared to be leakier than the stable inactivation observed in their eutherian counterparts (reviewed in Cooper et al., 1993). Therefore, there was a great interest in investigating the marsupial X chromosome and X inactivation in greater detail, a task in which marsupial BAC libraries have been indispensable.

The first step towards gaining a deeper understanding of X inactivation in marsupials was determining the gene content of the marsupial X chromosome. Early gene mapping studies showed that not all genes located on the human X chromosome were present on the X in marsupials. This was supported by cross-species chromosome painting which showed that the human X chromosome could be divided into two regions; one being a region conserved on the X chromosome both in marsupials and human, referred to as the X conserved region (XCR), and a region added to the X chromosome in the eutherian lineage - the X added region (XAR) (Glas et al., 1999; Wilcox et al., 1996). This added region corresponded to most of the short arm of the human X chromosome.

Progress in determining the boundaries of the XCR and XAR was slow until the release of the opossum genome assembly, which revealed this boundary in this species and pathed the way for detailed gene mapping in a second species, the tammar wallaby. Wallaby specific overgos were designed for human X-borne genes from sequence generated by the genome sequencing project and used to screen the wallaby BAC library in large pools. BACs for these genes were mapped to wallaby chromosomes using FISH. Genes from the XAR mapped to chromosome 5 (52 genes) and the XCR genes mapped to the X chromosome (47 genes). This mapping data enabled comparisons in gene order to be made between wallaby, opossum and human, revealing a surprising level of rearrangement on the X chromosome between these species (Deakin et al., 2008b).

One region that was of particular interest for comparative gene mapping in marsupials, given the differences in X inactivation between marsupials and eutherians, was the X inactivation center (XIC) located within the XCR on the human X chromosome. This region contains the *XIST* (X inactive specific transcript) gene, a master regulatory non-coding RNA transcribed from the inactive X, and a number of other non-coding RNAs that play an important role in X inactivation (reviewed in Avner & Heard, 2001). The *XIST* gene is poorly conserved between eutherian species (Chureau et al., 2002; Duret et al., 2006; Hendrich et al.,

1993; Nesterova et al., 2001). Sequence similarity searches failed to identify any sequence with homology to *XIST*. As a consequence, a BAC-based approach was taken to determine whether *XIST* was present in marsupials.

Three independent research teams used similar BAC-based approaches to determine the location of genes flanking the eutherian XIC locus on marsupial chromosomes. Shevchenko et al (2007) isolated BACs containing *XIST*-flanking genes as well as other genes from the XCR in two opossum species (*M.domestica* and *D.virginiana*). FISH-mapping of these BACs in both species revealed an evolutionary breakpoint between *XIST*-flanking genes. Likewise, Davidow et al (2007) and Hore et al (2007) mapped BACs identified to contain *XIST*-flanking genes from BAC-end sequence data generated as part of the opossum genome project and mapped them to different regions of the *M.domestica* X chromosome. Further sequence searches around these flanking genes failed to identify an orthologue of *XIST* (Davidow et al., 2007; Duret et al., 2006) and it was concluded that the *XIST* gene is absent in marsupials (Davidow et al., 2007; Hore et al., 2007). This conclusion was further supported by mapping of *XIST*-flanking genes to opposite ends of the tammar wallaby X chromosome (Deakin et al., 2008b). Hence, marsupial X inactivation is not under the control of *XIST* but then this raised more questions regarding marsupial X inactivation. Is there is a marsupial specific X inactivation centre? To answer this question, a more detailed investigation of the status of inactivation of marsupial X-borne genes was required.

Fortunately, the BACs isolated for mapping genes to the tammar wallaby X chromosome could be used construct an 'activity map' of the tammar wallaby X chromosome, where the inactivation status of X-borne genes at different locations along the X was determined. By using RNA-FISH, a technique that detects the nascent transcript, it was possible to determine the inactivation status of an X-borne gene within individual nuclei. The large insert size of BAC clones makes them ideal for hybridization and detection of the nascent transcript. Al Nadaf et al (2010) determined the inactivation status of 32 X-borne genes. As was suggested by earlier studies using isozymes, X inactivation in marsupials is incomplete. Every gene tested showed a percentage (5 - 68%) of cells with expression from both X chromosomes. This activity map of the wallaby X chromosome demonstrated no relationship between location on the X chromosome and extent of inactivation, suggesting that there is no polar spread of inactivation from a marsupial-specific inactivation center (Al Nadaf et al., 2010).

Although there are still many questions to be answered concerning marsupial X chromosome inactivation, BAC clones have proven to be extremely valuable resources for these studies and have resulted in the rapid advance of knowledge in this field. Further work is already underway to construct activity maps of genes in other species, using BACs from the opossum and the devil. Including a further species, the bandicoot (*I. macrourus*) would be particularly interesting as this species has an extreme version of X inactivation where they eliminate one sex chromosome (either a X in females or the Y in males) from somatic cells. The availability of a BAC library for this species makes it possible that this research could be carried out in the future.

2.1.2 Gene content of the marsupial Y chromosome

Although gene poor, the Y chromosome has an exceptionally important function, being responsible for sex determination and other functions in male sex and reproduction. A comparison of the chimpanzee and human Y chromosomes demonstrates the rapid evolution of the Y chromosome (Hughes et al., 2010). Extending this comparison to include marsupials would provide even further insight into the evolution of this remarkable chromosome. Orthologues of several eutherian Y-borne genes were mapped to the Y chromosome of marsupials but it was of more interest to see if there were novel genes found on the marsupial Y, which could be revealed by sequencing a marsupial Y chromosome.

Sequencing of the highly repetitive Y chromosome is extremely difficult by shot-gun sequencing. A BAC-based approach is seen as the best option to obtain well-assembled sequence. A novel method has been used to obtain Y specific BAC clones in the wallaby, in which the Y chromosome was isolated by flow sorting or manual microdissection and used to probe a wallaby BAC library and create a sub-library enriched with Y-specific BAC clones (Sankovic et al., 2006). Sequencing of two of these clones resulted in the identification of novel genes on the Y chromosome, *HUWE1Y* and *PHF6Y* (Sankovic et al., 2005). These genes are not on the Y chromosome of eutherians but do have a homologue on the X chromosome. It is hoped that more of these Y-specific BACs will be sequenced in the future to enable the evolutionary history of the therian Y chromosome to be unraveled.

2.1.3 Gene content of the platypus sex chromosomes

Monotremes, like other mammals, have male heteromorphic sex chromosomes, but their sex chromosome system is somewhat complex. Female platypuses have five different pairs of X chromosomes and their male counterparts have five X and five Y chromosomes that form a multivalent translocation chain during male meiosis (Grutzner et al., 2004). Similarly, the echidna (*T. aculeatus*) has five X chromosomes in females, and five X and four Y chromosomes in males (Rens et al., 2007). Early gene mapping studies using RISH with several heterologous probes suggested that at least one monotreme X chromosome shared homology with the therian X (Spencer et al., 1991; Watson et al., 1992; Watson et al., 1990). Subsequent mapping of BAC clones containing *XIST*-flanking genes indicated that at least some therian X-borne genes had an autosomal location in the platypus (Hore et al., 2007). The sequencing of the platypus genome made it possible to more thoroughly investigate the gene content of all platypus X chromosomes. By FISH-mapping BACs end-sequenced as part of the genome project, it became evident that, in contrast to the original gene mapping data, the platypus X chromosomes share no homology the therian X. Instead, at least some of the X chromosomes share homology with the chicken Z. Genes from the XCR were located on platypus chromosome 6 (Veyrunes et al., 2008). Furthermore, mapping of platypus X chromosome BACs onto male chromosomes identified the pseudoautosomal regions on the platypus Y chromosomes, providing the first glimpse into the gene content of the platypus Ys. Finding a lack of homology between monotreme and therian X chromosomes had a major impact on our understanding of the timing of therian sex chromosome evolution and provided surprising insight into the ancestral amniote sex determination system, which may have resembled the ZW system observed in birds (Waters & Marshall Graves, 2009).

The complicated sex chromosome system of monotremes makes determining the sequence of platypus Y chromosomes especially interesting. Since only a female platypus was sequenced as part of the genome project, no Y-specific sequence was obtained (Warren et al., 2008). Kortschak et al (2009) isolated and sequenced six Y-specific platypus BAC clones. The gene content of these BACs has not been reported but a detailed analysis of the repeat content has shown a bias towards the insertion of young SINE and LINE elements and segmental duplications (Kortschak et al., 2009). As some differences in gene content between platypus and echidna X chromosomes have been identified, a comparison of the gene and repeat content of their Y chromosomes could provide important insight into the evolution of this complicatied sex chromosome system. Undoubtedly, a BAC-based approach will continue to be the best strategy for obtaining Y-specific sequence.

The unexpected finding of no homology between monotreme and therian sex chromosomes begged the question as to how monotremes achieved dosage compensation. BAC clones were instrumental in determining the expression status of platypus X-borne genes in RNA-FISH experiments. Genes on platypus X chromosomes were monoallelically expressed in approximately 50% of cells and were biallelically expressed in the remainder, and so it appeared that the platypus employs a very leaky form of X inactivation for dosage compensation (Deakin et al., 2008a). This stochastic transcriptional regulation resembled the leaky inactivation of X-borne genes in the wallaby (Al Nadaf et al., 2010), suggesting that despite different origins of the X chromosome in monotremes and marsupials, their X inactivation mechanisms may have evolved from an ancient stochastic monoallelic expression mechanism that has subsequently independently evolved in the three major mammalian lineages (Deakin et al., 2008a, 2009).

In an attempt to further characterize features of the platypus X inactivation system, BAC clones were used to examine replication timing and X chromosome condensation, two features common to X inactivation in therian mammals. Replication timing of X-borne genes was determined by hybridizing fluorescently labeled BACs to interphase nuclei and counting the number of nuclei with asynchronous replication represented by double dots over one homologue of the gene of interest and a single dot over the other. These dot assays revealed asynchronous replication of some regions on the X chromosomes, namely those not shared on the Y (Ho et al., 2009). Condensation status of three platypus X chromosomes was determined by hybridizing two BACs mapped to opposite ends of the chromosome and measuring the distance between the two signals on the two X chromosome homologues. Only one X chromosome (X_3) displayed signs of differences in chromosome condensation. Consequently, chromosome condensation may not play a significant role in platypus dosage compensation (Ho et al., 2009). It would be interesting to perform these same experiments in echidna for comparative purposes. Since an echidna BAC library is available, it is hoped that this data will be obtained in the future and such a comparison made.

2.2 Evolution of genomic imprinting

Most autosomal genes in diploid organisms are expressed from both the maternal and paternal copies at equal levels. However, there are roughly 80 exceptional genes in eutherian mammals that are monoallelically expressed in a parent of origin fashion. The silent allele is marked (imprinted) by epigenetic features, such as CpG methylation and histone modifications. The evolution of a genomic imprinting mechanism appears counterintuitive since surely it would be more advantageous to have two expressed copies of a gene to protect the individual against deleterious mutations occurring in one copy. Consequently, genomic imprinting raises many questions regarding the how and why genomic imprinting evolved, although there appears to be some link between the evolution of viviparity and genomic imprinting (Hore et al., 2007).

By examining the orthologues of eutherian imprinted genes in marsupials and monotremes, it becomes possible to begin addressing some the questions regarding the evolution of genomic imprinting. Gene mapping with BAC probes and BAC clone sequencing have contributed greatly to research in this area. Below are just a few examples where the use of BAC clones has proven critical for tracing the evolutionary history of imprinted loci. Even in the fairly well covered opossum genome sequence, it has been necessary to sequence BAC clones spanning regions of interest in order to fill gaps in the genome assembly. Major conclusions drawn from these studies propose that imprinting arose independently at each imprinted locus and that the acquisition of imprinting involved changes to the genomic landscape of the imprinted region.

2.2.1 Analysis of the *IGF2/H19* locus

The *IGF2* imprinted locus has been extensively characterized in humans and mice, and was the first gene reported to be subject to genomic imprinting in marsupials (but not monotremes) (O'Neill et al., 2000). Elucidating the mechanism by which this is achieved was the subject of a number of subsequent studies. Sequence comparisons between the non-imprinted *IGF2* locus of platypus and the imprinted locus of marsupial and eutherian mammals were made in an attempt to identify potential sequence elements required for imprinting of this locus. A platypus BAC clone containing the *IGF2* gene was fully sequenced and compared to opossum, mouse and human. This study failed to identify any sites of differential methylation in intragenic regions but did uncover strong association of imprinting with both a lack of short interspersed transposable elements (SINEs) and an intragenic conserved inverted repeat (Weidman et al., 2004). Isolation of an opossum BAC clone (Lawton et al., 2007) and more extensive interrogation of the locus, identified a differentially methylated region (Lawton et al., 2008). This BAC clone was used in RNA-FISH experiments to show that demethylation of this differentially methylated region results in biallelic expression of *IGF2* (Lawton et al., 2008). Therefore, differential DNA methylation does indeed play a role in *IGF2* imprinting in marsupials.

In humans, *H19* is a maternally expressed long non-coding RNA located near the *IGF2* locus. While protein coding genes in this region were easily identified from genome sequence, the low level of sequence conservation typical of non-coding RNAs made the identification of *H19* more challenging. Three wallaby BACs spanning the the *IGF2/H19* locus were isolated by screening the library with probes designed from all available vertebrate sequences for genes within the region (Smits et al., 2008). Sensitive sequence similarity searches of the sequence obtained from these BAC clones identified a putative *H19* transcript with 51% identity to human *H19*. This sequence was found to be absent from the opossum genome assembly and hence, a BAC clone containing the opossum *H19* orthologue was isolated and sequenced. Like eutherians, *H19* is maternally expressed in marsupials (Smits et al., 2008).

2.2.2 Assembly of the Prader-Willi/Angelman's syndrome locus

Mutations in imprinted genes on human chromosome 15q11-q13 are responsible for the neurological disorders Prader-Willi and Angelman's syndrome. Imprinting of genes in this region is controlled by an imprinting control region (ICR) located within the Prader-Willi/Angelman's syndrome domain (Kantor et al., 2004). The ICR is flanked by the paternally expressed *SNRPN* gene and maternally expressed *UBE3A*. A cross-species comparison of the arrangement of these two genes across vertebrates uncovered an unexpected finding. A wallaby BAC clone containing the *SNRPN* gene mapped to wallaby chromosome 1, whereas the BAC containing the *UBE3A* localized to the short arm of chromosome 5. Furthermore, a fully sequenced platypus BAC clone containing *UBE3A* identified the gene adjacent to be *CNGA3*, a human chromosome 2 gene (Rapkins et al., 2006). Subsequent analysis of the chicken, zebrafish and opossum genome sequence assemblies unequivocally showed this to be the ancestral arrangement, with *UBE3A* adjacent to *CNG3A* while *SRNPN* is located elsewhere in the genome. Both *UBE3A* and *SRNPN* were found to be biallelic expressed in marsupials and monotremes. It appears that the other imprinted genes found in this region in eutherians do not exist in marsupials and originated from RNA copies of genes located in other parts of the genome. Rapkins et al (2006) concluded that these genes only became subject to genomic imprinting when the region was assembled in the eutherian lineage. This study also provided the first evidence that genomic imprinting was acquired by different loci at different times during mammalian evolution.

2.2.3 Evolution of the Callipyge imprinted locus

The Callipyge locus, so named after a muscle trait observed in sheep, contains a cluster of three paternally expressed genes (*DIO3, DLK1, RTL1*). In order to carry out a comprehensive analysis of this locus, seven platypus and 13 wallaby overlapping BAC clones were fully sequenced and assembled into a single contig for each species (Edwards et al., 2008). Comparative genome analysis revealed that the genomic landscape of this locus has undergone a number of changes during mammalian evolution. In marsupials, the locus is twice the size of the orthologous region in eutherians as a result of an accumulation of LINE1 repeats. In addition, there has been selection against SINE repeats in eutherians along with an increase in GC and CpG island content. Over 140 evolutionary conserved regions were found by phylogenetic footprinting but none of these regions corresponded to the imprint control element identified in eutherians. These findings were consistent with the absence of imprinted expression for this locus both in monotremes and marsupials. Similar to the situation described above for the Prader-Willi/Angelman locus, it appears that a retrotransposition event resulted in the formation of a novel gene in eutherians and it was suggested that this may have been the driving force behind the evolution of imprinting at this locus (Edwards et al., 2008).

2.3 Major Histocompatibility Complex

One the most studied regions of the vertebrate genome is the Major Histocompatibility Complex (MHC), a region central to the vertebrate immune response. In humans, the MHC is a large, gene dense region, spanning 3.6Mb and containing 224 genes divided into three regions; Class I, II and III (MHC Sequencing Consortium, 1999). Classes I and II encode genes involved in endogenous and exogenous antigen presentation respectively. Class III contains immune genes, involved in the inflammatory, complement and heat-shock responses, as well as a number of non-immune genes. This organization is in stark contrast to the chicken MHC consisting of only 19 genes within a 92kb region (Kaufman et al., 1999), making it difficult to establish the evolutionary history of the MHC. The position of marsupials and monotremes in vertebrate phylogeny ideally situates them to bridge the gap between chicken and eutherian mammal divergence and trace the evolutionary history of this important region. BAC clones have, once again, played an essential role in the study of the MHC organization and sequencing in marsupials and monotremes.

2.3.1 The opossum MHC

The opossum MHC was the first multi-megabase region to be annotated for the opossum genome project. Annotation of this region was performed on preliminary genome assemblies MonDom1 and MonDom2. The MHC region in MonDom1 was distributed across five sequence scaffolds. Previous mapping localized MHC Class I genes to different locations on opossum chromosome 2, with genes *UB* and *UC* located at the telomeric end of the short arm (Belov et al., 2006) and *UA1* located near the centromere on the long of arm (Gouin et al., 2006). Thus, it was imperative that this assembly of MHC scaffolds was accurately determined to establish whether the separation of these genes was the result of a chromosomal rearrangement or a transposition event. This was achieved by isolating BAC clones corresponding to the ends of the MHC scaffolds. All BACs from these scaffolds, with the exception of one containing *UB* and *UC,* mapped to the centromeric region of chromosome 2. As a result of this information, the MHC was assembled into a single scaffold in the MonDom2 assembly (Belov et al., 2006). Furthermore, mapping of BAC clones from the genes at either end of this large scaffold enabled the orientation of the MHC on the chromosomes to be determined.

The complete annotation of this region provided the necessary information required to start piecing together the changes which have occurred throughout vertebrate evolution. In contrast to the chicken, the MHC of the opossum spans almost 4Mb and contains at least 140 genes, making it similar in size and complexity to the human MHC (Belov et al., 2006). However, the opossum has a very different gene organization with Class I and II genes found interspersed rather than separated by the Class III region as they are in eutherian mammals. This organization is similar to that of other vertebrates, such a shark and frog, suggesting that the marsupial organization is similar to that of the vertebrate ancestor and the eutherian organization is derived.

2.3.2 Mapping and sequencing of the wallaby MHC

The opossum and wallaby are distantly related species, having diverged from a common ancestor around 60 - 80 mya, making a comparison of these two species similar to the informative human-mouse comparison. Unlike the opossum, the wallaby genome was only lightly sequenced (Renfree et al., 2011), leaving many gaps in the genome assembly. If detailed comparative analysis was to be carried out on the MHC, an alternative approach was required.

Initial comparative analysis of these two MHCs was carried out using gene mapping. BACs containing MHC genes from all three Classes were isolated from a tammar wallaby BAC library. These clones were FISH-mapped to wallaby chromosomes with startling results. All Class II and Class III genes, as well as MHC flanking genes, mapped to the expected location on chromosome 2. Surprisingly, all of the MHC Class I BACs mapped to locations on every chromosome except chromosome 2 and the sex chromosomes (Deakin et al., 2007). This unexpected and unprecedented result made a more thorough analysis of these genes critical. As a result, a concerted effort was made to sequence the entire tammar wallaby MHC, including the 'core' MHC located on chromosome 2 and many of the dispersed Class I genes found elsewhere in the genome. A BAC-based approach was taken, with the idea of constructing a BAC-contig across the core MHC, as well as sequencing the dispersed Class I genes.

After finding Class I genes dispersed across the genome, a thorough screening of the wallaby BAC library was performed in order to isolate as many Class I genes as possible. As a result four additional BAC clones containing Class I genes were isolated, with FISH-mapping of these BACs localizing these genes to the core MHC region on chromosome 2 (Siddle et al., 2009). Complete sequencing of these BACs identified six Class I genes within the core MHC, which were interspersed with antigen processing genes and a Class II gene. Sequencing of ten BACs mapping outside this region identified nine Class I genes with open reading frames. In depth sequence analysis of these BACs revealed a tendency for Kangaroo Endogenous Retroviral Element (KERV) to flank these dispersed Class I genes, suggesting that this element may be implicated in the movement of these genes to regions outside the core MHC (Siddle et al., 2009).

A BAC contig across the core MHC on wallaby chromosome 2 was constructed for sequencing purposes (Siddle et al., 2011). Unfortunately, despite extensive library screening with overgo probes designed from BAC end sequence, a single contig spanning the entire region was not obtained. Instead, the isolated BACs assembled into nine contigs and three 'orphaned' BACs. The order of these contigs and orphaned BACs was determined using BAC clones as probes for FISH on metaphase chromosome spreads and interphase nuclei. The resulting 4.7Mb sequence contained 129 predicted genes from all three MHC Classes. A comparison of the gene arrangement between wallaby, opossum and other vertebrates indicated that the wallaby MHC has a novel MHC gene arrangement, even within the core MHC. The wallaby Class II genes have undergone an expansion, residing in two clusters either side of the Class III region. Once again, KERV sequences are prominent in this region and may have contributed to the overall genomic instability of the wallaby MHC region (Siddle et al., 2011).

2.3.3 The MHC in monotremes

Although the platypus genome has been sequenced, the high GC and repeat content hampered this sequencing effort, leaving the assembly with many more gaps than other mammalian genomes sequenced to a six-fold depth by Sanger sequencing (Warren et al., 2008). As a result, complete annotation of the platypus MHC as a region was impossible because MHC genes were found on many sequence contigs and/or scaffolds. However, three BAC clones were completely sequenced and mapped to platypus chromosomes (Dohm et al., 2007). One of these BACs, localized to chromosome 3, only contained a processed class I pseudogene. Of the remaining two BACs, one contained two Class I genes and two Class II genes as well as antigen processing genes, while the other contained mainly Class III genes. The most surprising result came from FISH-mapping, which revealed that platypus MHC is not contiguous and maps to the pseudoautosomal region of two pairs of sex chromsomes. The Class I and II genes were located on X_3/Y_3 and the Class III region on X_4/Y_4. Subsequent FISH-mapping of BACs containing these same genes in the echidna demonstrated that this separation of the MHC onto two different pairs of sex chromosomes was a common feature for monotremes. Monotremes are the only mammals known to date to have the MHC reside on sex chromosomes (Dohm et al., 2007).

2.4 Tracing the evolutionary history of globin genes

Haemoglobin is essential for oxygen transportation in vertebrates. The haemoglobin molecule is encoded by members of the α- and β-globin gene clusters. These gene clusters were presumed to have arisen from a single globin gene that duplicated to form a combined α- and β-globin gene cluster as is seen in amphibians (Jeffreys et al., 1980). It was proposed that either a fission event or a chromosome duplication event, followed by independent evolution of the duplicate copies, gave rise to the separate α- and β-globin gene clusters observed in amniotes (Jeffreys et al., 1980). Determining the gene content of the marsupial and monotreme globin gene clusters has had a tremendous impact in this field. This work was facilitated by sequencing and mapping of BACs containing globin genes.

The discovery of a novel β-like globin gene called *HBW* residing adjacent to the wallaby α-globin cluster provided support for the chromosome duplication hypothesis (Wheeler et al., 2004). Further support was provided when BAC clones from the dunnart (*S.macroura*) spanning the separate α- and β-globin gene clusters were sequenced and it was found that, like the wallaby, the *HBW* was adjacent to the α-globin cluster (De Leo et al., 2005). The next obvious step in testing the chromosome duplication hypothesis was to determine the organization of the platypus α- and β-globin gene clusters. The fragmented nature of the platypus genome meant that a BAC-based approach was required to obtain a more complete sequence of the alpha and beta globin gene clusters in this species (Patel et al., 2008). Analysis of the sequence obtained from these BAC clones was instrumental in the formation of a new hypothesis for the evolution of these gene clusters.

The platypus α-globin cluster also contained a copy of *HBW*, which taken on its own would support the chromosome duplication hypothesis. However, an examination of the genes flanking the two clusters revealed that the combined α/β-globin cluster in amphibians was flanked by the same genes as the α-globin cluster in all amniotes, whereas the β-globin cluster in amniotes was surrounded by olfactory receptors. This led to a hypothesis where the α-globin cluster in amniotes was proposed to correspond to the original α/β-globin cluster present in other vertebrates. The β-globin cluster was proposed to have evolved after a copy of the original β-globin gene (*HBW*) was transposed into an array of olfactory receptors (Patel et al., 2008).

3. Anchoring marsupial and monotreme genome assemblies

Genome sequence data on its own is an extremely valuable resource but it is also equally as important to know how the genome fits together. BACs have played an essential role in anchoring marsupial and monotreme sequence to chromosomes. Different approaches have been taken that have utilized BACs to improve genome assemblies, with the strategy employed dependent the quality of the genome assembly.

The opossum and platypus genome projects employed BACs in a similar fashion. BAC-end sequencing was used to assist in connecting sequence contigs into scaffolds (Mikkelsen et al., 2007; Warren et al., 2008). Scaffolds were anchored and oriented on chromosomes by FISH-mapping BACs from ends of sequence scaffolds (Duke et al., 2007; Warren et al., 2008). For the opossum genome, the mapping of 381 BACs resulted in 97% of the genome being assigned to chromosomes (Duke et al., 2007). The more fragmented nature of the platypus genome assembly made it more difficult to anchor each scaffold but FISH-mapping of BACs assigned 198 scaffolds, corresponding to approximately 20% of the genome, to chromosomes (Warren et al., 2008).

Anchoring of the even more fragmented wallaby and devil genome assemblies required a different approach. A novel approach was developed to anchor the low-coverage wallaby genome sequence to chromosomes. A cytogenetic map of the genome was constructed by mapping BACs containing genes from the ends of human-opossum conserved gene blocks. This strategy was first trialed on tammar wallaby chromosome 5 (Deakin et al., 2008b) and later applied to the entire genome (Renfree et al., 2011). A virtual map of the wallaby genome was made by extrapolating from the content of these mapped conserved blocks from the opossum genome assembly, thereby allowing the location of each gene on tammar wallaby chromosomes to be predicted (Wang et al., 2011a). A similar approach is being used to construct a map of the devil genome, which has been sequenced entirely by next generation sequencing (Miller et al., 2011).

4. BACs and marsupial linkage maps

Linkage (genetic) maps are a useful resource as they provide information not only on the order of genetic markers on a chromosome but on the location and frequencies of crossover events. Such maps are even more valuable if the maps are anchored to chromosomes and integrated with available genome assembly and/or cytogenetic mapping data. Linkage maps have been constructed for two marsupial species, opossum (Samollow et al., 2007) and the wallaby (Wang et al., 2011b). BACs containing markers at the ends of linkage groups have been used to cytogenetically assign these groups to chromosomes and determine the genome coverage of the linkage maps (Samollow et al., 2007; Wang et al., 2011b). The opossum linkage map was integrated with the genome assembly and cytogenetic map by FISH-mapping 34 BAC clones from the ends of linkage groups (Duke et al., 2007; Samollow et al., 2007). A sophisticated approach was used in the marker selection for construction of the wallaby linkage map to facilitate the integration of cytogenetic and linkage map data. Three strategies were developed to fill gaps in the 1st generation linkage map (Zenger et al., 2002) using information from BACs. The first strategy involved identifying microsatellites in BACs that had been previously assigned to chromosomes by FISH. The second strategy identified microsatellites within BAC end sequences and the third used the wallaby genome sequence to identify microsatellite markers near BACs that had been mapped by FISH. This resulted in a linkage map that could easily be incorporated with the physical map data to generate an integrated map (Wang et al., 2011a, 2011b). Information from the integrated map has been used to improve and anchor the tammar wallaby genome assembly (Renfree et al., 2011).

5. Conclusions

Our understanding of marsupial and monotreme genomes has been greatly advanced due in large part to the availability of BAC libraries for several key species. With the emergence of Devil Facial Tumour Disease (DFTD), a transmissible cancer threatening the extinction of this species in the wild within the next 25 years (McCallum et al., 2007), many marsupial researchers are are focusing their efforts on characterization of this devastating disease. BAC libraries are playing a play a major role in this work, building on the strategies developed for other species described here to rapidly gain as much information as possible on the normal and DFTD tumour genomes. In addition, genome sequencing of other marsupial and monotremes species is currently underway using next generation sequencing technology and it is anticipated that BAC libraries will continue to be a very precious resource for comparative genomic studies in these species.

6. References

Al Nadaf, S., Waters, P. D., Koina, E., Deakin, J. E., Jordan, K. S. & Graves, J. A. (2010). Activity map of the tammar X chromosome shows that marsupial X inactivation is incomplete and escape is stochastic. *Genome Biology,* Vol.11, No.12, pp. R122, ISSN 1465-6914

Avner, P. & Heard, E. (2001). X-chromosome inactivation: counting, choice and initiation. *Nature Review Genetics,* Vol.2, No.1, pp. 59-67, ISSN 1471-0056

Belov, K., Deakin, J. E., Papenfuss, A. T., Baker, M. L., Melman, S. D., Siddle, H. V., Gouin, N., Goode, D. L., Sargeant, T. J., Robinson, M. D., Wakefield, M. J., Mahony, S., Cross, J. G., Benos, P. V., Samollow, P. B., Speed, T. P., Graves, J. A. & Miller, R. D. (2006). Reconstructing an ancestral mammalian immune supercomplex from a marsupial major histocompatibility complex. *PLoS Biology,* Vol.4, No.3, pp. e46, ISSN 1545-7885

Bininda-Emonds, O. R., Cardillo, M., Jones, K. E., MacPhee, R. D., Beck, R. M., Grenyer, R., Price, S. A., Vos, R. A., Gittleman, J. L. & Purvis, A. (2007). The delayed rise of present-day mammals. *Nature,* Vol.446, No.7135, pp. 507-512, ISSN 1476-4687

Chapman, M. A., Charchar, F. J., Kinston, S., Bird, C. P., Grafham, D., Rogers, J., Grutzner, F., Graves, J. A., Green, A. R. & Gottgens, B. (2003). Comparative and functional analyses of LYL1 loci establish marsupial sequences as a model for phylogenetic footprinting. *Genomics,* Vol.81, No.3, pp. 249-259, ISSN 0888-7543

Chureau, C., Prissette, M., Bourdet, A., Barbe, V., Cattolico, L., Jones, L., Eggen, A., Avner, P. & Duret, L. (2002). Comparative sequence analysis of the X-inactivation center region in mouse, human, and bovine. *Genome Research,* Vol.12, No.6, pp. 894-908, ISSN 1088-9051

Cooper, D. W., Johnston, P. G. & Graves, J. A. M. (1993). X-inactivation in marsupials and monotremes. *Seminars in Developmental Biology,* Vol.4, No.2, pp. 117-128, ISSN 1044-5781

Davidow, L. S., Breen, M., Duke, S. E., Samollow, P. B., McCarrey, J. R. & Lee, J. T. (2007). The search for a marsupial XIC reveals a break with vertebrate synteny. *Chromosome Research,* Vol.15, No.2, pp.137-146, ISSN 0967-3849

De Leo, A. A., Wheeler, D., Lefevre, C., Cheng, J. F., Hope, R., Kuliwaba, J., Nicholas, K. R., Westerman, M. & Graves, J. A. (2005). Sequencing and mapping hemoglobin gene clusters in the Australian model dasyurid marsupial *Sminthopsis macroura. Cytogenetic and Genome Research,* Vol.108, No.4, pp. 333-341, ISSN 1424-859X

Deakin, J. E., Olp, J. J., Graves, J. A. & Miller, R. D. (2006a). Physical mapping of immunoglobulin loci IGH@, IGK@, and IGL@ in the opossum (*Monodelphis domestica*). *Cytogenetic and Genome Research,* Vol.114, No.1, pp. 94H, ISSN 1424-859X

Deakin, J. E., Parra, Z. E., Graves, J. A. & Miller, R. D. (2006b). Physical mapping of T cell receptor loci (TRA@, TRB@, TRD@ and TRG@) in the opossum (*Monodelphis domestica*). *Cytogenetic and Genome Research,* Vol.112, No.3-4, pp .342K, ISSN 1424-859X

Deakin, J. E., Siddle, H. V., Cross, J. G., Belov, K. & Graves, J. A. (2007). Class I genes have split from the MHC in the tammar wallaby. *Cytogenetic and Genome Research,* Vol.116, No.3, pp. 205-211, ISSN 1424-859X

Deakin, J. E., Hore, T. A., Koina, E. & Marshall Graves, J. A. (2008a). The status of dosage compensation in the multiple X chromosomes of the platypus. *PLoS Genetics,* Vol.4, No.7, pp. e1000140, ISSN 1553-7404

Deakin, J. E., Koina, E., Waters, P. D., Doherty, R., Patel, V. S., Delbridge, M. L., Dobson, B., Fong, J., Hu, Y., van den Hurk, C., Pask, A. J., Shaw, G., Smith, C., Thompson, K., Wakefield, M. J., Yu, H., Renfree, M. B. & Graves, J. A. (2008b). Physical map of two tammar wallaby chromosomes: a strategy for mapping in non-model mammals. *Chromosome Research,* Vol.16, No.8, pp. 1159-1175, ISSN 1573-6849

Deakin, J. E., Chaumeil, J., Hore, T. A. & Marshall Graves, J. A. (2009). Unravelling the evolutionary origins of X chromosome inactivation in mammals: insights from marsupials and monotremes. *Chromosome Research,* Vol.17, No.5, pp. 671-685, ISSN 1573-6849

Deakin, J. E. (2010). Physical and Comparative Gene Maps in Marsupials. In: *Marsupial Genetics and Genomics,* Deakin, J. E., Waters, P. D. & Graves, J. A. M. (Eds), 101-115, Springer, ISBN 978-90-481-9022-5, Dordrecht, Heidelberg, London, New York

Dohm, J. C., Tsend-Ayush, E., Reinhardt, R., Grutzner, F. & Himmelbauer, H. (2007). Disruption and pseudoautosomal localization of the major histocompatibility complex in monotremes. *Genome Biology,* Vol.8, No.8, pp. R175, ISSN 1465-6914

Duke, S. E., Samollow, P. B., Mauceli, E., Lindblad-Toh, K. & Breen, M. (2007). Integrated cytogenetic BAC map of the genome of the gray, short-tailed opossum, Monodelphis domestica. *Chromosome Research,* Vol.15, No.3, pp. 361-370, ISSN 0967-3849

Duret, L., Chureau, C., Samain, S., Weissenbach, J. & Avner, P. (2006). The Xist RNA gene evolved in eutherians by pseudogenization of a protein-coding gene. *Science,* Vol.312, No.5780, pp. 1653-1655, ISSN 0036-8075

Edwards, C. A., Mungall, A. J., Matthews, L., Ryder, E., Gray, D. J., Pask, A. J., Shaw, G., Graves, J. A., Rogers, J., Dunham, I., Renfree, M. B. & Ferguson-Smith, A. C. (2008). The evolution of the DLK1-DIO3 imprinted domain in mammals. *PLoS Biology,* Vol.6, No.6, pp.e135, ISSN 1545-7885

Edwards, M. J., Hinds, L. A., Deane, E. M. & Deakin, J. E. (2011). Physical Mapping of Innate Immune Genes, Mucins and Lysozymes, and Other Non-Mucin Proteins in the Tammar Wallaby (*Macropus eugenii*). *Cytogenetic and Genome Research,* ISSN 1424-859X

El-Mogharbel, N., Wakefield, M., Deakin, J. E., Tsend-Ayush, E., Grutzner, F., Alsop, A., Ezaz, T. & Marshall Graves, J. A. (2007). DMRT gene cluster analysis in the platypus: new insights into genomic organization and regulatory regions. *Genomics,* Vol.89, No.1, pp. 10-21, ISSN 0888-7543

Glas, R., Marshall Graves, J. A., Toder, R., Ferguson-Smith, M. & O'Brien, P. C. (1999). Cross-species chromosome painting between human and marsupial directly demonstrates the ancient region of the mammalian X. *Mammalian Genome,* Vol.10, No.11, pp. 1115-1116, ISSN 0938-8990

Gouin, N., Deakin, J. E., Miska, K. B., Miller, R. D., Kammerer, C. M., Graves, J. A., VandeBerg, J. L. & Samollow, P. B. (2006). Linkage mapping and physical localization of the major histocompatibility complex region of the marsupial

Monodelphis domestica. Cytogenetic and Genome Research, Vol.112, No.3-4, pp. 277-285, ISSN 1424-859X

Grafodatskaya, D., Rens, W., Wallis, M. C., Trifonov, V., O'Brien, P. C., Clarke, O., Graves, J. A. & Ferguson-Smith, M. A. (2007). Search for the sex-determining switch in monotremes: mapping WT1, SF1, LHX1, LHX2, FGF9, WNT4, RSPO1 and GATA4 in platypus. *Chromosome Research,* Vol.15, No.6, pp.777-785, ISSN 0967-3849

Grutzner, F., Rens, W., Tsend-Ayush, E., El-Mogharbel, N., O'Brien, P. C. M., Jones, R. C., Ferguson-Smith, M. A. & Graves, J. A. M. (2004). In the platypus a meiotic chain of ten sex chromosomes shares genes with the bird Z and mammal X chromosomes. *Nature,* Vol.432, No.7019, pp. 913-917, ISSN 0028-0836

Hayman, D. L. & Martin, P. G. (1965). Sex chromosome mosaicism in the marsupial genera *Isoodon* and *Perameles*. *Genetics,* Vol.52, No.6, pp. 1201-1206, ISSN 0016-6731

Hendrich, B. D., Brown, C. J. & Willard, H. F. (1993). Evolutionary conservation of possible functional domains of the human and murine XIST genes. *Human Molecular Genetics,* Vol.2, No.6, pp. 663-672, ISSN 0964-6906

Ho, K. K., Deakin, J. E., Wright, M. L., Graves, J. A. & Grutzner, F. (2009). Replication asynchrony and differential condensation of X chromosomes in female platypus (*Ornithorhynchus anatinus*). *Reproduction Fertility and Development,* Vol.21, No.8, pp. 952-963, ISSN 1031-3613

Hope, R. M. & Cooper, D. W. (1990). Marsupial and monotreme breeding in wild and captive populations: towards a laboratory marsupial. In: *Mammals from Pouches and Eggs: Genetics, Breeding and Evolution of Marsupials and Monotremes,* Graves, J. A. M., Hope, R. M. & Cooper, D. W. (Eds), CSIRO Press, ISBN 978-06-430-5020-4, Melbourne

Hore, T. A., Koina, E., Wakefield, M. J. & Marshall Graves, J. A. (2007). The region homologous to the X-chromosome inactivation centre has been disrupted in marsupial and monotreme mammals. *Chromosome Research,* Vol.15, No.2, pp. 147-161, ISSN 0967-3849

Hore, T.A., Rapkins, R.W., Graves, J.A.M. (2007). Construction and evolution of imprinted loci in mammals. *Trends in Genetics,* Vol.23., No.9., pp. 440-448, ISSN 0168-9525

Hughes, J.F., Skaletsky, H., Pyntikova, T., Graves,T.A., van Daalen, S.K., Minx, P.J., Fulton, R.S., McGrath, S.D., Locke, D.P., Friedman, C., Trask, B.J., Mardis, E.R., Warren, W.C., Repping, S., Rozen, S., Wilson, R.K., Page, D.C. (2010). Chimpanzee and human Y chromosomes are remarkably divergent in structure and gene content. *Nature,* Vol.463, No.7280, pp. 536-539, ISSN 0028-0836

Jeffreys, A. J., Wilson, V., Wood, D., Simons, J. P., Kay, R. M. & Williams, J. G. (1980). Linkage of adult alpha- and beta-globin genes in *X. laevis* and gene duplication by tetraploidization. *Cell* Vol.21, No.2, pp. 555-564, ISSN 0092-8674

Johnston, P. G., Watson, C. M., Adams, M. & Paull, D. J. (2002). Sex chromosome elimination, X chromosome inactivation and reactivation in the southern brown bandicoot *Isoodon obesulus* (Marsupialia: Peramelidae). *Cytogenetic and Genome Research,* Vol.99, No.1-4, pp.119-124, ISSN 1424-859X

Kantor, B., Makedonski, K., Green-Finberg, Y., Shemer, R. & Razin, A. (2004). Control elements within the PWS/AS imprinting box and their function in the imprinting process. *Human Molecular Genetics,* Vol.13, No.7, pp. 751-762, ISSN 0964-6906

Kaufman, J., Milne, S., Gobel, T. W., Walker, B. A., Jacob, J. P., Auffray, C., Zoorob, R. & Beck, S. (1999). The chicken B locus is a minimal essential major histocompatibility complex. *Nature,* Vol.401, No.6756, pp. 923-925, ISSN 0028-0836

Koina, E., Wakefield, M. J., Walcher, C., Disteche, C. M., Whitehead, S., Ross, M. & Marshall Graves, J. A. (2005). Isolation, X location and activity of the marsupial homologue of SLC16A2, an XIST-flanking gene in eutherian mammals. *Chromosome Research,* Vol.13, No.7, pp.687-698, ISSN 0967-3849

Kortschak, R. D., Tsend-Ayush, E. & Grutzner, F. (2009). Analysis of SINE and LINE repeat content of Y chromosomes in the platypus, *Ornithorhynchus anatinus. Reproduction Fertility and Development,* Vol.21, No.8, pp. 964-975, ISSN 1031-3613

Lawton, B. R., Obergfell, C., O'Neill, R. J. & O'Neill, M. J. (2007). Physical mapping of the IGF2 locus in the South American opossum Monodelphis domestica. *Cytogenetic and Genome Research,* Vol.116, No.1-2, pp.130-131, 1424-859X

Lawton, B. R., Carone, B. R., Obergfell, C. J., Ferreri, G. C., Gondolphi, C. M., Vandeberg, J. L., Imumorin, I., O'Neill, R. J. & O'Neill, M. J. (2008). Genomic imprinting of IGF2 in marsupials is methylation dependent. *BMC Genomics,* Vol.9, pp.205, ISSN 1471-2164

McCallum, H., Tompkins, D. M., Jones, M., Lachish, S., Marvanek, S., Lazenby, B., Hocking, G., Wiersma, J. & Hawkins, C. E. (2007). Distribution and impacts of Tasmanian devil facial tumor disease. *Ecohealth,* Vol.4, No.3, pp.318-325, ISSN 1612-9202

MHC Sequencing Consortium (1999). Complete sequence and gene map of a human major histocompatibility complex. *Nature,* Vol.401, No.6756, pp. 921-923, ISSN 0028-0836

Mikkelsen, T. S., Wakefield, M. J., Aken, B., Amemiya, C. T., Chang, J. L., Duke, S., Garber, M., Gentles, A. J., Goodstadt, L., Heger, A., Jurka, J., Kamal, M., Mauceli, E., Searle, S. M., Sharpe, T., Baker, M. L., Batzer, M. A., Benos, P. V., Belov, K., Clamp, M., Cook, A., Cuff, J., Das, R., Davidow, L., Deakin, J. E., Fazzari, M. J., Glass, J. L., Grabherr, M., Greally, J. M., Gu, W., Hore, T. A., Huttley, G. A., Kleber, M., Jirtle, R. L., Koina, E., Lee, J. T., Mahony, S., Marra, M. A., Miller, R. D., Nicholls, R. D., Oda, M., Papenfuss, A. T., Parra, Z. E., Pollock, D. D., Ray, D. A., Schein, J. E., Speed, T. P., Thompson, K., VandeBerg, J. L., Wade, C. M., Walker, J. A., Waters, P. D., Webber, C., Weidman, J. R., Xie, X., Zody, M. C., Graves, J. A., Ponting, C. P., Breen, M., Samollow, P. B., Lander, E. S. & Lindblad-Toh, K. (2007). Genome of the marsupial Monodelphis domestica reveals innovation in non-coding sequences. *Nature,* Vol.447, No.7141, pp. 167-177, ISSN 1476-4687

Miller, W., Hayes, V. M., Ratan, A., Petersen, D. C., Wittekindt, N. E., Miller, J., Walenz, B., Knight, J., Qi, J., Zhao, F., Wang, Q., Bedoya-Reina, O. C., Katiyar, N., Tomsho, L. P., Kasson, L. M., Hardie, R. A., Woodbridge, P., Tindall, E. A., Bertelsen, M. F., Dixon, D., Pyecroft, S., Helgen, K. M., Lesk, A. M., Pringle, T. H., Patterson, N., Zhang, Y., Kreiss, A., Woods, G. M., Jones, M. E. & Schuster, S. C. (2011). Genetic diversity and population structure of the endangered marsupial *Sarcophilus harrisii* (Tasmanian devil). *Proceeding of the National Academy of Science U S A,* Vol.108, No.30, pp. 12348-12353, ISSN 1091-6490

Murchison, E. P., Tovar, C., Hsu, A., Bender, H. S., Kheradpour, P., Rebbeck, C. A., Obendorf, D., Conlan, C., Bahlo, M., Blizzard, C. A., Pyecroft, S., Kreiss, A., Kellis, M., Stark, A., Harkins, T. T., Marshall Graves, J. A., Woods, G. M., Hannon, G. J. & Papenfuss, A. T. (2010). The Tasmanian devil transcriptome reveals Schwann cell origins of a clonally transmissible cancer. *Science,* Vol.327, No.5961, pp. 84-87, ISSN 1095-9203

Nesterova, T. B., Slobodyanyuk, S. Y., Elisaphenko, E. A., Shevchenko, A. I., Johnston, C., Pavlova, M. E., Rogozin, I. B., Kolesnikov, N. N., Brockdorff, N. & Zakian, S. M. (2001). Characterization of the genomic Xist locus in rodents reveals conservation of overall gene structure and tandem repeats but rapid evolution of unique sequence. *Genome Research,* Vol.11, No.5, pp. 833-849, ISSN 1088-9051

O'Neill, M. J., Ingram, R. S., Vrana, P. B. & Tilghman, S. M. (2000). Allelic expression of IGF2 in marsupials and birds. *Development Genes and Evolution,* Vol.210, No.1, pp. 18-20, ISSN 0949-944X

Patel, V. S., Cooper, S. J., Deakin, J. E., Fulton, B., Graves, T., Warren, W. C., Wilson, R. K. & Graves, J. A. (2008). Platypus globin genes and flanking loci suggest a new insertional model for beta-globin evolution in birds and mammals. *BMC Biology,* Vol.6, pp.34, ISSN 1741-7007

Premzl, M., Delbridge, M., Gready, J. E., Wilson, P., Johnson, M., Davis, J., Kuczek, E. & Marshall Graves, J. A. (2005). The prion protein gene: identifying regulatory signals using marsupial sequence. *Gene,* Vol.349, pp.121-134, ISSN 0378-1119

Rapkins, R. W., Hore, T., Smithwick, M., Ager, E., Pask, A. J., Renfree, M. B., Kohn, M., Hameister, H., Nicholls, R. D., Deakin, J. E. & Graves, J. A. (2006). Recent assembly of an imprinted domain from non-imprinted components. *PLoS Genetics,* Vol.2, No.10, pp. e182, ISSN 1553-7404

Renfree, M. B., Papenfuss, A. T., Deakin, J. E., Lindsay, J., Heider, T., Belov, K., Rens, W., Waters, P. D., Pharo, E. A., Shaw, G., Wong, E. S., Lefevre, C. M., Nicholas, K. R., Kuroki, Y., Wakefield, M. J., Zenger, K. R., Wang, C., Ferguson-Smith, M., Nicholas, F. W., Hickford, D., Yu, H., Short, K. R., Siddle, H. V., Frankenberg, S. R., Chew, K. Y., Menzies, B. R., Stringer, J. M., Suzuki, S., Hore, T. A., Delbridge, M. L., Mohammadi, A., Schneider, N. Y., Hu, Y., O'Hara, W., Al Nadaf, S., Wu, C., Feng, Z. P., Cocks, B. G., Wang, J., Flicek, P., Searle, S. M., Fairley, S., Beal, K., Herrero, J., Carone, D. M., Suzuki, Y., Sagano, S., Toyoda, A., Sakaki, Y., Kondo, S., Nishida, Y., Tatsumoto, S., Mandiou, I., Hsu, A., McColl, K. A., Landsell, B., Weinstock, G., Kuczek, E., McGrath, A., Wilson, P., Men, A., Hazar-Rethinam, M., Hall, A., Davies, J., Wood, D., Williams, S., Sundaravadanam, Y., Muzny, D. M., Jhangiani, S. N., Lewis, L. R., Morgan, M. B., Okwuonu, G. O., Ruiz, S. J., Santibanez, J., Nazareth, L., Cree, A., Fowler, G., Kovar, C. L., Dinh, H. H., Joshi, V., Jing, C., Lara, F., Thornton, R., Chen, L., Deng, J., Liu, Y., Shen, J. Y., Song, X. Z., Edson, J., Troon, C., Thomas, D., Stephens, A., Yapa, L., Levchenko, T., Gibbs, R. A., Cooper, D. W., Speed, T. P., Fujiyama, A., Graves, J. A., O'Neill, R. J., Pask, A. J., Forrest, S. M. & Worley, K. C. (2011). Genome sequence of an Australian kangaroo, Macropus eugenii, provides insight into the evolution of mammalian reproduction and development. *Genome Biology,* Vol.12, No.8, pp. R81, ISSN 1465-6914

Rens, W., O'Brien, P. C. M., Grutzner, F., Clarke, O., Graphodatskaya, D., Tsend-Ayush, E., Trifonov, V. A., Skelton, H., Wallis, M. C., Johnston, S., Veyrunes, F., Graves, J. A. M. & Ferguson-Smith, M. A. (2007). The multiple sex chromosomes of platypus and echidna are not completely identical and several share homology with the avian Z. *Genome Biology,* Vol.8, No.11, pp. R243, ISSN 1474-760X

Samollow, P. B. (2006). Status and applications of genomic resources for the gray, short-tailed opossum. *Australian Journal of Zoology,* Vol.54, No.3, pp. 173-196, ISSN 0004-959X

Samollow, P. B., Gouin, N., Miethke, P., Mahaney, S. M., Kenney, M., VandeBerg, J. L., Graves, J. A. & Kammerer, C. M. (2007). A microsatellite-based, physically anchored linkage map for the gray, short-tailed opossum (*Monodelphis domestica*). *Chromosome Research,* Vol.15, No.3, pp. 269-281, ISSN 0967-3849

Sanderson, C. E., Belov, K. & Deakin, J. E. (2009). Physical mapping of immune genes in the tammar wallaby (*Macropus eugenii*). *Cytogenetic and Genome Research,* Vol.127, No.1, pp. 21-25, ISSN 1424-859X

Sankovic, N., Bawden, W., Martyn, J., Graves, J. A. M. & Zuelke, K. (2005). Construction of a marsupial bacterial artificial chromosome library from the model Australian marsupial, the tammar wallaby (*Macropus eugenii*). *Australian Journal of Zoology,* Vol.53, No.6, pp. 389-393, ISSN 0004-959X

Sankovic, N., Delbridge, M. L., Grutzner, F., Ferguson-Smith, M. A., O'Brien, P. C. & Marshall Graves, J. A. (2006). Construction of a highly enriched marsupial Y chromosome-specific BAC sub-library using isolated Y chromosomes. *Chromosome Research,* Vol.14, No.6, pp. 657-664, 0967-3849

Siddle, H. V., Deakin, J. E., Coggill, P., Hart, E., Cheng, Y., Wong, E. S., Harrow, J., Beck, S. & Belov, K. (2009). MHC-linked and un-linked class I genes in the wallaby. *BMC Genomics,* Vol.10, pp. 310, ISSN 1471-2164

Siddle, H. V., Deakin, J. E., Coggill, P., Wilming, L. G., Harrow, J., Kaufman, J., Beck, S. & Belov, K. (2011). The tammar wallaby major histocompatibility complex shows evidence of past genomic instability. *BMC Genomics,* Vol.12, No.1, pp. 421, ISSN 1471-2164

Smits, G., Mungall, A. J., Griffiths-Jones, S., Smith, P., Beury, D., Matthews, L., Rogers, J., Pask, A. J., Shaw, G., VandeBerg, J. L., McCarrey, J. R., Renfree, M. B., Reik, W. & Dunham, I. (2008). Conservation of the H19 noncoding RNA and H19-IGF2 imprinting mechanism in therians. *Nature Genetics,* Vol.40, No.8, pp. 971-976, ISSN 1546-1718

Spencer, J. A., Watson, J. M., Lubahn, D. B., Joseph, D. R., French, F. S., Wilson, E. M. & Graves, J. A. (1991). The androgen receptor gene is located on a highly conserved region of the X chromosomes of marsupial and monotreme as well as eutherian mammals. *Journal of Heredity,* Vol.82, No.2, pp. 134-139, ISSN 0022-1503

Veyrunes, F., Waters, P. D., Miethke, P., Rens, W., McMillan, D., Alsop, A. E., Grutzner, F., Deakin, J. E., Whittington, C. M., Schatzkamer, K., Kremitzki, C. L., Graves, T., Ferguson-Smith, M. A., Warren, W. & Marshall Graves, J. A. (2008). Bird-like sex chromosomes of platypus imply recent origin of mammal sex chromosomes. *Genome Research,* Vol.18, No.6, pp.965-973, ISSN 1088-9051

Wakefield, M. J. & Alsop, A. E. (2006). Assignment of BReast Cancer Associated 1 (BRCA1) to tammar wallaby (*Macropus eugenii*) chromosome 2q3 by in situ hybridization. *Cytogenetic and Genome Research,* Vol.112, No.1-2, pp. 180C, ISSN 1424-859X

Wakefield, M. J., Anderson, M., Chang, E., Wei, K. J., Kaul, R., Graves, J. A., Grutzner, F. & Deeb, S. S. (2008). Cone visual pigments of monotremes: filling the phylogenetic gap. *Visual Neuroscience,* Vol.25, No.3, pp. 257-264, ISSN 1469-8714

Wallis, M. C., Delbridge, M. L., Pask, A. J., Alsop, A. E., Grutzner, F., O'Brien, P. C., Rens, W., Ferguson-Smith, M. A. & Graves, J. A. (2007a). Mapping platypus SOX genes; autosomal location of SOX9 excludes it from sex determining role. *Cytogenetic and Genome Research,* Vol.116, No.3, pp. 232-234, ISSN 1424-859X

Wallis, M. C., Waters, P. D., Delbridge, M. L., Kirby, P. J., Pask, A. J., Grutzner, F., Rens, W., Ferguson-Smith, M. A. & Graves, J. A. (2007b). Sex determination in platypus and echidna: autosomal location of SOX3 confirms the absence of SRY from monotremes. *Chromosome Research*, Vol.15, No.8, pp. 949-959, ISSN 0967-3849

Wang, C., Deakin, J. E., Rens, W., Zenger, K. R., Belov, K., Marshall Graves, J. A. & Nicholas, F. W. (2011a). A first-generation integrated tammar wallaby map and its use in creating a tammar wallaby first-generation virtual genome map. *BMC Genomics*, Vol.12, pp. 422, ISSN 1471-2164

Wang, C., Webley, L., Wei, K. J., Wakefield, M. J., Patel, H. R., Deakin, J. E., Alsop, A., Graves, J. A., Cooper, D. W., Nicholas, F. W. & Zenger, K. R. (2011b). A second-generation anchored genetic linkage map of the tammar wallaby (*Macropus eugenii*). *BMC Genetics*, Vol.12, No.1, pp. 72, ISSN 1471-2156

Warren, W. C., Hillier, L. W., Marshall Graves, J. A., Birney, E., Ponting, C. P., Grutzner, F., Belov, K., Miller, W., Clarke, L., Chinwalla, A. T., Yang, S. P., Heger, A., Locke, D. P., Miethke, P., Waters, P. D., Veyrunes, F., Fulton, L., Fulton, B., Graves, T., Wallis, J., Puente, X. S., Lopez-Otin, C., Ordonez, G. R., Eichler, E. E., Chen, L., Cheng, Z., Deakin, J. E., Alsop, A., Thompson, K., Kirby, P., Papenfuss, A. T., Wakefield, M. J., Olender, T., Lancet, D., Huttley, G. A., Smit, A. F., Pask, A., Temple-Smith, P., Batzer, M. A., Walker, J. A., Konkel, M. K., Harris, R. S., Whittington, C. M., Wong, E. S., Gemmell, N. J., Buschiazzo, E., Vargas Jentzsch, I. M., Merkel, A., Schmitz, J., Zemann, A., Churakov, G., Kriegs, J. O., Brosius, J., Murchison, E. P., Sachidanandam, R., Smith, C., Hannon, G. J., Tsend-Ayush, E., McMillan, D., Attenborough, R., Rens, W., Ferguson-Smith, M., Lefevre, C. M., Sharp, J. A., Nicholas, K. R., Ray, D. A., Kube, M., Reinhardt, R., Pringle, T. H., Taylor, J., Jones, R. C., Nixon, B., Dacheux, J. L., Niwa, H., Sekita, Y., Huang, X., Stark, A., Kheradpour, P., Kellis, M., Flicek, P., Chen, Y., Webber, C., Hardison, R., Nelson, J., Hallsworth-Pepin, K., Delehaunty, K., Markovic, C., Minx, P., Feng, Y., Kremitzki, C., Mitreva, M., Glasscock, J., Wylie, T., Wohldmann, P., Thiru, P., Nhan, M. N., Pohl, C. S., Smith, S. M., Hou, S., Nefedov, M., de Jong, P. J., Renfree, M. B., Mardis, E. R. & Wilson, R. K. (2008). Genome analysis of the platypus reveals unique signatures of evolution. *Nature*, Vol.453, No.7192, pp. 175-183, ISSN 1476-4687

Waters, P. D. & Marshall Graves, J. A. (2009). Monotreme sex chromosomes--implications for the evolution of amniote sex chromosomes. *Reproduction Fertility and Development*, Vol.21, No.8, pp. 943-951, 1031-3613

Watson, J. M., Spencer, J. A., Riggs, A. D. & Graves, J. A. (1990). The X chromosome of monotremes shares a highly conserved region with the eutherian and marsupial X chromosomes despite the absence of X chromosome inactivation. *Proceedings of the National Academy of Sciences U S A*, Vol.87, No.18, pp. 7125-7129, ISSN 0027-8424

Watson, J. M., Riggs, A. & Graves, J. A. (1992). Gene mapping studies confirm the homology between the platypus X and echidna X1 chromosomes and identify a conserved ancestral monotreme X chromosome. *Chromosoma*, Vol.101, No.10, pp. 596-601, ISSN 0009-5915

Weidman, J. R., Murphy, S. K., Nolan, C. M., Dietrich, F. S. & Jirtle, R. L. (2004). Phylogenetic footprint analysis of IGF2 in extant mammals. *Genome Research*, Vol.14, No.9, pp. 1726-1732, ISSN 1088-9051

Wheeler, D., Hope, R. M., Cooper, S. J., Gooley, A. A. & Holland, R. A. (2004). Linkage of the beta-like omega-globin gene to alpha-like globin genes in an Australian marsupial supports the chromosome duplication model for separation of globin gene clusters. *Journal of Molecular Evolution,* Vol.58, No.6, pp. 642-652, ISSN 0022-2844

Whittington, C. M., Papenfuss, A. T., Bansal, P., Torres, A. M., Wong, E. S., Deakin, J. E., Graves, T., Alsop, A., Schatzkamer, K., Kremitzki, C., Ponting, C. P., Temple-Smith, P., Warren, W. C., Kuchel, P. W. & Belov, K. (2008). Defensins and the convergent evolution of platypus and reptile venom genes. *Genome Research,* Vol.18, No.6, pp. 986-994, ISSN 1088-9051

Wilcox, S. A., Watson, J. M., Spencer, J. A. & Graves, J. A. (1996). Comparative mapping identifies the fusion point of an ancient mammalian X-autosomal rearrangement. *Genomics,* Vol.35, No.1, pp. 66-70, ISSN 0888-7543

Zenger, K. R., McKenzie, L. M. & Cooper, D. W. (2002). The first comprehensive genetic linkage map of a marsupial: the tammar wallaby (*Macropus eugenii*). *Genetics,* Vol.162, No.1, pp.321-330, ISSN 0016-6731

2

Recombineering of BAC DNA for the Generation of Transgenic Mice

John J. Armstrong and Karen K. Hirschi
Yale Cardiovascular Research Center
Yale University School of Medicine, New Haven, CT
USA

1. Introduction

Bacterial Artificial Chromosomes (BAC) are low copy plasmids that stably maintain genomic DNA sequences hundreds of kilobases (Kb) in length. Thus, BAC plasmids usually contain the entire locus of one or more genes, enabling their use for genetic engineering and for the creation of genomic libraries for large-scale gene sequencing projects.

The use of BAC plasmids for transgenic gene expression is also gaining popularity over traditional proximal promoter driven transgene expression because the BAC typically contain most, if not all, of the important regulatory elements required to recapitulate endogenous gene expression (Giraldo and Montoliu 2001). Maintaining the coding sequence within its regulatory locus insulates the transgene from integration position dependent effects on expression enforced by nearby enhancers and heterochromatin (Wilson, Bellen et al. 1990).

Harnessing the power of BAC plasmids requires developing and optimizing methods for manipulation of the gene loci within the BAC, generally referred to as "Recombineering". This chapter will discuss the use of such technology to modify BAC DNA, specifically for the introduction of a fluorescent reporter to mark the expression of a gene of interest. We will also discuss the characterization of BAC transgenic mice and their experimental utility.

2. Overview of recombineering technology

Traditional cloning approaches rely on the presence of unique restriction enzyme sites for modification of plasmid DNA via a series of digestions and ligations to incorporate or remove desired DNA sequences. Unfortunately, most restriction enzyme sequences are not unique or conveniently located within the genomic sequence. Thus, the availability/use of restriction sites is often a limiting factor when attempting to modify plasmid DNA using such approaches.

Recombineering technology achieves DNA modification using a phage homologous recombination system, which uses linear DNA as template. Thus, an investigator can use linear targeting vectors containing 5′ and 3′ arms with homology to a target locus to introduce new DNA sequence. Subtle changes can now be achieved, including single point mutations. Since the modification to the locus is based solely on sequence present in the target and not restriction enzymes, DNA can be introduced to the target wherever needed. Thus, recombineering technology has opened up an unlimited number of possibilities for genetic modifications.

Prior to the use of lambda phage in Recombineering, the study of homologous recombination in E. coli laid the groundwork for the use of this technology. Homologous recombination via linear DNA is suppressed in E. coli by the recBCD enzyme complex. In the recBCD model, the enzyme complex moves destructively along double strand breaks. The recB and recC subunits operate as helicases while recD operates as an exonuclease. Thus, in recBCD wild type E. coli strains, recombination does not occur through linear DNA, as reviewed elsewhere (Myers and Stahl 1994; Yeeles and Dillingham 2010).

The recBCD complex moves destructively along linear DNA until it encounters a DNA motif called a chi site. The chi site motif is a "recombination hotspot" that facilitates homologous recombination by ejecting the recD subunit responsible for the exonuclease activity of the complex, but does not affect the helicase activity. The helicase activity results in single strand DNA that serves as substrate for homologous recombination. This molecular reaction was exploited by incorporation of chi sites into targeting vectors to enable homologous recombination of target genes.

Cloning by homologous recombination was also studied in recBCD deficient strains. These experiments used recBCD mutants in an attempt to modify bacterial chromosomal and plasmid DNA with a linear DNA targeting construct (Jasin and Schimmel 1984; Oliner, Kinzler et al. 1993). The success of these studies demonstrated the effectiveness of cloning by homologous recombination, but was dependent on the use of specialized bacterial strains with constitutively active recombination enzymes. This enzyme activity resulted in unwanted intramolecular rearrangements in the modified plasmid (Copeland, Jenkins et al. 2001). Therefore, the recBCD mutant strains were limited in their use in cloning by homologous recombination.

The Recombineering technology used today is based on the lambda phage Red double strand break repair system, which uses the phage proteins *exo*, *bet*, and *gam*. This system is initiated when the 5′to3′ exonuclease *exo* digests linear double stranded (ds) DNA leaving a 3′ overhang of single stranded (ss) DNA. The resultant 3′ssDNA is coated by the *bet* protein, which facilitates its annealing to a complementary strand of DNA. Once the homologous DNA is annealed, the 3′OH becomes a priming site for DNA replication resulting in double strand break repair.

The activities of *exo*, *bet*, and *gam* have been adapted for BAC cloning by homologous recombination. The dsDNA substrate for *exo* is a linear targeting construct with 5′ and 3′ homology to the target locus. The linear targeting construct can be generated by PCR or excised from a plasmid. The resultant 3′ssDNA is coated by the *bet* protein and facilitates the annealing of the 3′ssDNA of the targeting construct to the targeted sequence on the BAC containing the gene locus. The linear targeting construct is unaffected by recBCD activity due to the presence of the *gam* protein, which inhibits recBCD binding to the dsDNA targeting construct (Murphy 2007).

Studies that employed the introduction of lambda phage Red double strand break repair into E. coli demonstrated that it was an efficient system for cloning by homologous recombination (Murphy 1998). Other studies demonstrated the use of homology arms as short as 27nt in length facilitated cloning by homologous recombination, with increasing efficiency with increased homology length. This study also investigated insert length between homologous arms and found it useful for inserts from 0-3100bp in length (Zhang, Buchholz et al. 1998). It was with these findings that lambda phage Red double strand break repair was adopted for cloning by homologous recombination.

3. Recombineering materials for BAC modification

There are currently two different resources for phage based Recombineering tools, Genebridges and NCI Frederick Systems. The major differences between the systems are how the phage genes are introduced into bacteria and what types of gene modifications they are capable of.

3.1 Genebridges plasmid based system

Genebridges (http://www.genebridges.com/) has adopted a plasmid based system to introduce the phage recombination proteins under a pBAD promoter (Noll, Hampp et al. 2009). In this system, expression of the lambda phage genes is repressed by araC dimer. Gene expression is induced when the araC dimer is released from the pBAD promoter in the presence of L-arabinose. This system introduces the recombineering apparatus into the E. coli strain containing the BAC and does not require specialized bacterial strains that endogenously express the recombination proteins.

3.2 NCI Frederick bacterial based system

The National Cancer Institute http:// web.ncifcrf.gov/ research/ brb/ recombineering Information.aspx offers E. coli strains that contain the phage recombination proteins stably integrated into the genome. The proteins are under transcriptional control of the λPl promoter in concert with the temperature sensitive cl857 repressor. Transcription of the lambda phage proteins is repressed at 32°C. Repression of the λ Pl promoter is released by incubating cells at 42°C for 15 minutes. Placing the stably integrated phage proteins under tight transcriptional control circumvents the problems of unwanted recombination associated with constitutive expression (Warming, Costantino et al. 2005). NCI offers strains with an L-arabinose inducible Cre or Flpe expression, which are useful for excision of the selectable antibiotic markers used for positive selection of recombinants.

In addition, NCI has developed plasmids to facilitate the cloning of sequence from BAC DNA into high copy plasmids for downstream use as targeting vectors (Liu, Jenkins et al. 2003). Plasmids for epitope tagging new proteins of interest have also been developed, allowing the investigator to localize and purify proteins. This application speeds the functional characterization of proteins by eliminating the lengthy process of generating antibodies that would serve similar functions (Poser, Sarov et al. 2008).

4. Recombineering methods and applications for BAC modification

In this section, we will discuss the protocols and approaches used, in conjunction with NCI recombineering E. coli strains, for the modification of a BAC plasmid, which was used for the creation of a new mouse reporter line.

4.1 Experimental design

In our studies, we were interested in generating a reporter mouse line to monitor the expression of Smooth Muscle-α-Actin (SMA) in vivo and track the fate of cells expressing this gene. Transcriptional control of the SMA locus had been well defined in previous studies (Mack and Owens 1999; Mack, Thompson et al. 2000). Using that information, we decided to replace exon 2 (the first coding exon of SMA) and 50bp distal of the 3′ end of exon 2, with a myristoylated mCherry fluorescent reporter (Shaner, Campbell et al. 2004). Our rationale for this design was to a) insert the reporter in a manner where the first codon of the reporter replaced the first codon of SMA, and b) to avoid generating a fusion protein by removing the exon 2 splice site donor (Fig. 1a).

Our targeting construct consisted of a mCherry expression sequence with a myristoylation sequence upstream of a PGK-neomycin resistance cassette (Fig. 1a). The PGK-neomycin resistance cassette was flanked at the 5′ and 3′ end by FRT recombination sites. Inclusion of the PGK-neomycin resistance cassette facilitated selection of positive recombinant clones by kanamycin (Fig. 1b). Inclusion of the FRT recombination sites allowed for subsequent excision of the PGK-neomycin resistance cassette by Flpe recombinase, thus eliminating any unwanted transcription effects of the PGK promoter on the reporter (Fig. 1c). The targeting construct ended with a SV40 polyadenylation sequence to enable efficient transcription (Fig. 1a).

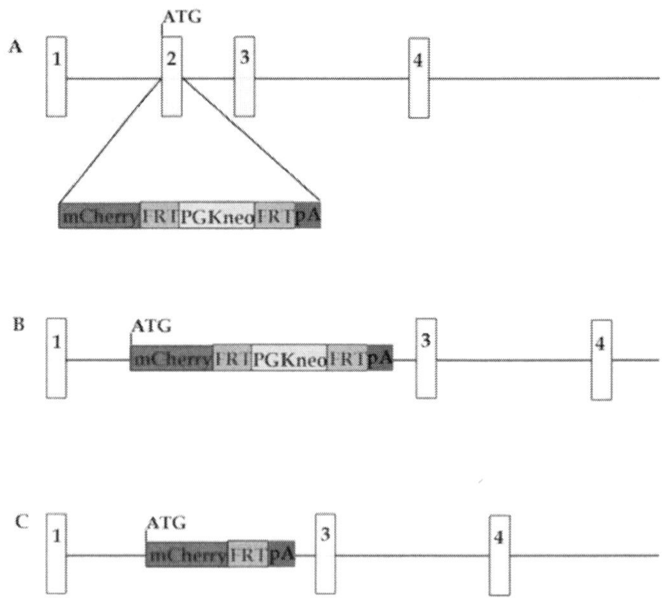

Fig. 1. Experimental design for the generation of a Smooth-Muscle-α-Actin (SMA) reporter construct. (a) The endogenous SMA locus and linear targeting construct. (b) The SMA locus correctly targeted by recombineering. (c) The final reporter construct following excision of PGK-neomycin selection cassette.

Our immunohistochemistry on cultured cells selected for mCherry expression by Fluorescent Activated Cell Sorting (FACS) showed a distinctly different membrane localization of our mCherry reporter compared to the cytoskeletal staining pattern of SMA indicating we had avoided generating a SMA-mCherry fusion (Armstrong, Larina et al. 2010).

4.2 Isolation of BAC DNA

Isolation of BAC DNA can be achieved using a standard alkaline lysis phenol/chloroform extraction followed by alcohol precipitation. However, these types of BAC DNA preparations are subject to contamination by genomic DNA. Column purification of BAC DNA, such as that achieved using NucleoBond BAC 100 kit by Clontech, will yield more pure and less degraded BAC DNA. With either procedure, the removal of fragmented linear genomic DNA can be achieved using ExoV exonuclease or increasing lysis of the bacterial wall using lysozyme, as needed. However, large BAC DNA sequences are susceptible to shearing, so one must take care during preparation so as not to degrade the intact BAC DNA.

Quick BAC prep

Although there is a considerable genomic DNA contamination using this Quick BAC Prep procedure, we found it to be sufficient for isolating BAC DNA for introduction into the SW105 recombineering bacterial strain.

1. Pick colony and place in 5ml LB with appropriate selection reagent and place in shaking incubator at proper temperature (30°C for SW bacterial series; 37°C for DH10 bacteria) overnight.
2. Pellet bacteria by centrifugation at 4,000xg for 15min.
3. Resuspend pellet in 1000ul of P1 buffer (50 mM Tris-HCl, pH 8.0, 10 mM EDTA) with 100ug/ml RNaseA.
4. Add 1000ul P2 buffer (200 mM NaOH, 1% SDS) for lysis reaction for 5 min at RT.
5. Add 1000ul P3 buffer (3.0 M potassium acetate, pH 5.5 adjusted with glacial acetic acid) for cell debris precipitation. Do not vortex! Vortexing BAC DNA will destroy it. Swirl to mix, and then incubate on ice for 5 min.
6. Centrifuge the prep at 4,000xg for 45 min at 4°C. Decant supernatant.
7. Precipitate BAC DNA with 2 ml 2-propanol, incubated at -20°C for 30min or more.
8. Centrifuge at 4,000xg for 30 min at 4°C to pellet DNA.
9. Decant supernatant and wash with 1 ml 70% ethanol. Centrifuge 4,000xg for 15 min at 4°C.
10. Decant supernatant and dry DNA pellet, but do not allow to completely dry. BAC DNA that has completely dried is difficult to resuspend.
11. Resuspend pellet in 200 ul TE at 4°C overnight. This prep should yield 500 ng-1 ug/ul DNA with approximately 50-75% genomic contamination.

4.3 Electroporation of BAC DNA into bacteria

Electroporation of bacteria is usually performed with an exponential decay electroporation instrument, although square wave instruments can be adapted for bacterial electroporation.

We found that for the initial electroporation of BAC DNA into recombineering bacterial strains, a "dirty" prep of BAC DNA containing genomic DNA prepared by alkaline lysis alcohol precipitation was sufficient.

With "Quick Prep" BAC DNA (described above), electroporation can be performed at 1.8KV with a time constant of 5µs. Cells are then resuspended in 1ml LB media and incubated for 1hr at 30°C. As with standard practice, 100ul cells are streaked on selection plates and incubated at 30°C overnight. To the remaining electroporated cells, 4ml LB medium are added and incubated overnight at 30°C. The following day, cells were plated on selection plates and transformed cells are obtained from this selection.

4.4 Preparation of induced electrocompetent cells

Electrocompetent cells are prepared by inoculation of 50 ml LB plus appropriate reagent to select for the BAC clone of interest at 1:50 dilution from an overnight culture. Cells are incubated at 30°C until the culture achieves an OD600 of 0.50, which is usually 3-4 hrs. Cells are then divided into two aliquots, one for induction of lambda phage proteins and one un-induced control.

To induce the lambda phage proteins, cells are incubated at 42°C for 15 minutes, and then harvested by centrifugation at 5000xg for 15 minutes at 4°C. Cells are then washed 3 times in 25ml ice-cold 0.2µm-filtered ddH_2O containing 10% glycerol, resuspended in 30µl 10% glycerol in ddH_2O, and then transferred to cooled cuvettes for electroporation.

4.5 Generating a targeting vector

Recombineering modifies the genomic locus using a linear targeting construct. The linear targeting construct can be generated either by PCR or excision from plasmid DNA, as discussed below.

4.5.1 PCR based approach

A convenient method for generating linear targeting vectors is via a PCR based approach in which primers are designed against two important sequence elements. The 5′ end of the primer contains sequence homology to the targeted locus while the 3′ end contains sequence for PCR amplification of the targeting construct. When these primers are used in a PCR reaction, the product produced contains both homology to the target and the targeting construct. Using a PCR based approach allows an investigator to use the same template to target an alternative locus. By changing the 5′ homology of the primers to match an alternative locus, a new linear targeting construct can be generated. Therefore, the template used for generating the linear targeting construct is modular.

4.5.2 Plasmid based approach

Cloning longer (100-500 bp) homology arms into the targeting construct is an alternative approach to generating a linear targeting vector. Using this type of construct is also useful for cloning sequence from BAC DNA into high copy plasmids for subsequent modification and use for in vitro cell targeting (i.e. embryonic stem cells). The linear targeting construct is released from the vector by restriction enzyme digest and gel purified. This approach reportedly reduces base changes that are sometimes generated during PCR.

4.6 Preparation and electroporation of targeting vector for BAC modification

In our studies, we generated a linear targeting vector by PCR. Following PCR, the reaction is DpnI digested to remove the plasmid template. This step is incorporated in an attempt to reduce background and false positive clones that would result from electroporation of plasmid template. The PCR product is then gel purified and used to electroporate cells containing the BAC clone. Cells are electroporated with 1ug of linear targeting vector using the parameters previously described in section 4.3. Recombinants are obtained by plating of the overnight culture, as described above.

4.7 Selection of positive recombinants

Selection of positive recombinants can be facilitated by inclusion of a PGK-neomycin selection cassette in the targeting vector that confers kanamycin resistance to recombinant clones. Cells are plated on LB agar plates containing 15ug/ml kanamycin and grown at 30°C overnight, as described above. Following antibiotic selection, clones are screened for correct targeting using DNA isolated from kanamycin resistant clones in a PCR reaction using two different sets of primers that would amplify a product across the 5′ and 3′ recombination sites. PCR reactions are then analyzed on 1% agarose gel to correctly targeted clones. The PCR products are then sequenced to ensure the recombination had taken place as expected. Southern Blot analysis is another method for identifying the inclusion of the targeting construct into the BAC. While Southern Blot is sensitive, it gives no information on the sequence and requires additional time and labeling reagents (i.e. radioactive nucleotides).

4.7.1 Identifying false positive recombinants

In most cases, the template DNA used in the PCR reaction generating the linear targeting vector is eliminated by a DpnI digest, followed by gel purification. However, despite using DpnI to digest methylated DNA used as template in the generation of linear targeting vector by PCR, some plasmid template may be introduced into bacteria during the electroporation (Fig. 2). Identification of this contamination can be facilitated by gel electrophoresis of uncut BAC DNA isolated from kanamycin resistant clones. Even though the linear targeting construct is gel purified, the plasmid template can transform kanamycin resistant clones. For example, the undigested supercoiled plasmid DNA and the linear targeting vector can run at approximately the same molecular weight, thus contaminating the linear targeting vector during gel purification, as occurred in our studies.

5. Characterization of BAC transgenic mice

5.1 BAC reporter expression recapitulates endogenous expression

To verify that the transgenic BAC reporter is expressed in the same temporal and spatial pattern as the protein of interest, immunohistochemistry is employed to evaluate tissue and cellular specificity. Not only is it important to demonstrate that the reporter and the protein of interest are co-expressed, but it is also important to demonstrate that the reporter does not exhibit ectopic expression, beyond endogenous expression. In situ hybridization can also be utilized to characterize the co-expression of the reporter and targeted gene of interest.

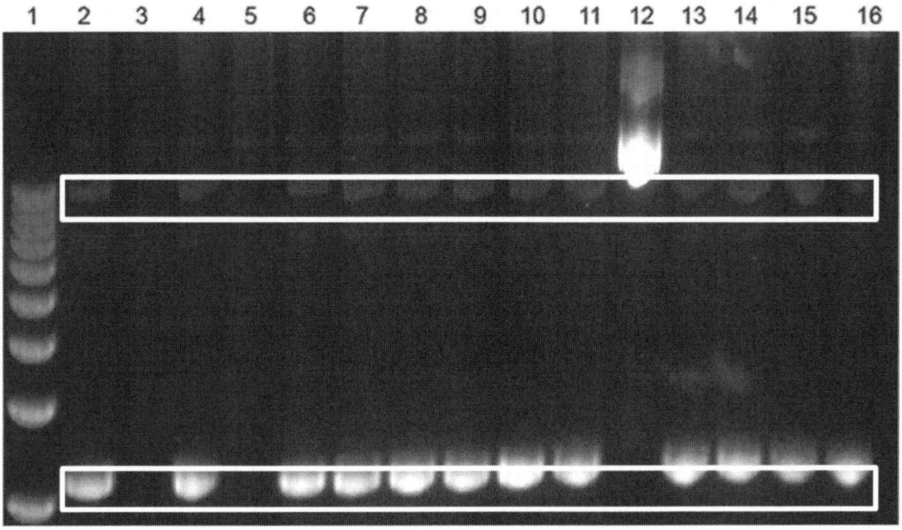

Fig. 2. Identification of PCR template contamination in selected BAC clones.
Uncut BAC DNA preps run on 1% agarose gel. BAC DNA is in the upper box, and contaminating PCR template in the lower box. Lane 1 - Hi Mark Ladder. Lanes 3, 4, and 6 are true positive recombinants without contaminating PCR template used to generate the linear targeting construct.

5.2 BAC transgene copy number

An assay was developed by Chandler and coworkers (Chandler, Chandler et al. 2007) to determine the number of copies of BAC transgene integrated into the genome. We adopted this method to measure copy number in our SMA-mCherry transgenic lines (Armstrong, Larina et al. 2010), and found that the level of expression of the reporter correlated with BAC copy number, which is useful for the comparative evaluation of the mouse lines.

5.3 Integration site of BAC transgene

Chandler and coworkers were also able to identify multiple integration sites in BAC reporters. This was achieved by employing the BAC transgene copy number assay described above and analyzing outcrossed F2 generations for the number of transgene copies (Chandler, Chandler et al. 2007). Since the level of reporter expression is dependent on transgene copy number, it is useful to identify founders that transmit a single integration site, ensuring consistent reporter expression in future progeny.

6. Utility of BAC transgenics

6.1 FACS analysis and isolation of live cells based on intracellular protein expression

One advantage of introducing a reporter into the gene locus of an intracellular protein is that one can then use the reporter mouse or cell line to identify and isolate live cells by Fluorescence Activated Cell Sorting (FACS) for in vitro applications. For example, we used the SMA-mCherry mice that we developed to isolate live cells expressing SMA from tissues during embryonic development (Fig. 3a), and demonstrated that the cells retain reporter expression after culture in vitro (Fig. 3b).

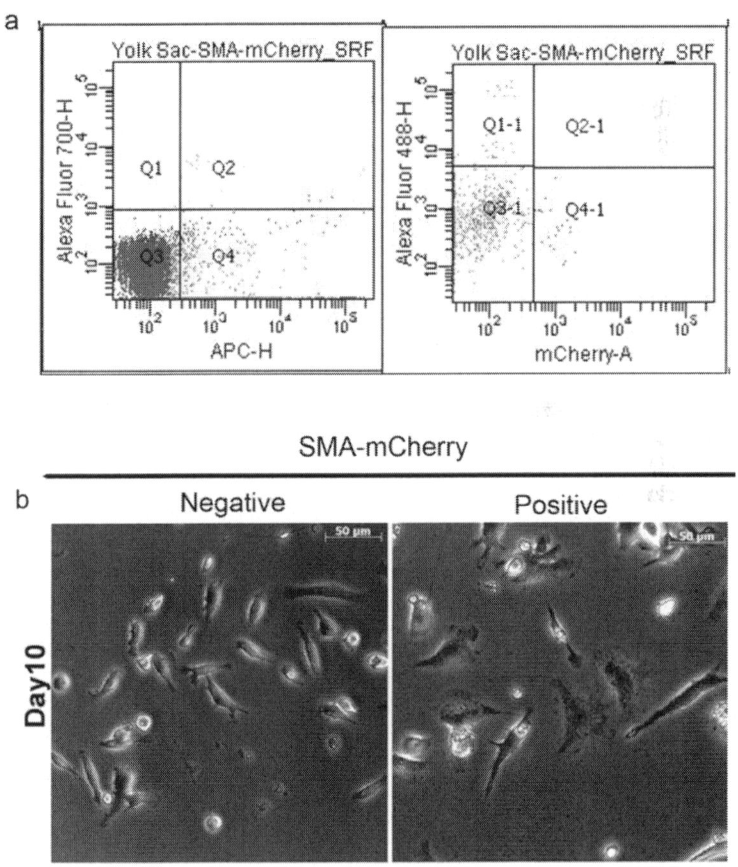

Fig. 3. Isolation of SMA positive cells by flow cytometry. (a) FACS analysis of E9.5 yolk sac endothelium reveals that a subset of endothelium expressed the mesenchymal marker SMA. (b) Expression of SMA was retained when the cells were cultured in vitro.

6.2 *In vivo* imaging of embryonic vascular development

BAC transgenic mice expressing reporter constructs can also be used to monitor and measure the dynamic emergence and fate of distinct cell types. For example, in our studies,

we adopted an embryo culture system previously used for in vivo imaging of the cardiovascular system and hemodynamics (Garcia, Udan et al. 2011; Garcia, Udan et al. 2011). We used the SMA-mCherry reporter mice, crossed to an endothelial specific Flk1-YFP reporter mouse line, to monitor endothelial-mesenchymal interactions during vascular development (Fraser, Hadjantonakis et al. 2005). Still images of a time-lapse experiment are shown in Figure 4. At early time points in the experiment (~E8.5), only (YFP+) endothelial cells are present within the developing yolk sac (Fig. 4a,d). As vascular development progresses during embryo culture (~E9.0), SMA expressing cells appear (Fig. 4b,e, arrowhead); their migration within the tissue can be monitored over time (Fig. 4c,f, arrowhead). Heart development and function can also be monitored and measured using these mice, as previously reported (Armstrong et al. 2010).

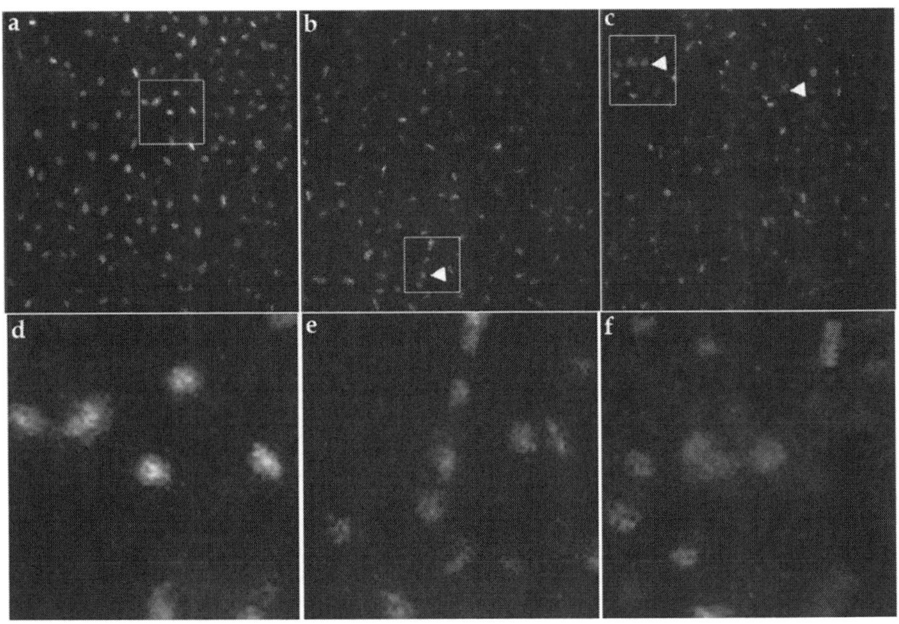

Fig. 4. Still series from in vivo imaging. (a,d) YFP+ endothelial cells prior to the emergence of SMA+ cells. At later time points, SMA-expressing cells beginning to appear (b,e), and their movement within the tissue can be monitored over time (c,d).

7. Conclusions

In summary, BAC clones provide a stable source of starting material for gene targeting. BAC clones, used in conjunction with Recombineering technology, provide investigators with numerous gene modifications approaches such as those needed to create transgenic reporter mouse lines. In addition to BAC transgenics, Recombineering technology can be employed for the construction of conditional alleles, point mutations, insertions and deletions. Thus, this technology is versatile and powerful.

8. References

Armstrong, J. J., I. V. Larina, et al. (2010). "Characterization of bacterial artificial chromosome transgenic mice expressing mCherry fluorescent protein substituted for the murine smooth muscle alpha-actin gene." Genesis 48(7): 457-463.

Chandler, K. J., R. L. Chandler, et al. (2007). "Relevance of BAC transgene copy number in mice: transgene copy number variation across multiple transgenic lines and correlations with transgene integrity and expression." Mamm Genome 18(10): 693-708.

Fraser, S. T., A. K. Hadjantonakis, et al. (2005). "Using a histone yellow fluorescent protein fusion for tagging and tracking endothelial cells in ES cells and mice." Genesis 42(3): 162-171.

Garcia, M. D., R. S. Udan, et al. (2011). "Live Imaging of Mouse Embryos." Cold Spring Harb Protoc 2011(4): pdb.top104-.

Garcia, M. D., R. S. Udan, et al. (2011). "Time-Lapse Imaging of Postimplantation Mouse Embryos." Cold Spring Harb Protoc 2011(4): pdb.prot5595-.

Giraldo, P. and L. Montoliu (2001). "Size matters: use of YACs, BACs and PACs in transgenic animals." Transgenic Res 10(2): 83-103.

Liu, P., N. A. Jenkins, et al. (2003). "A highly efficient recombineering-based method for generating conditional knockout mutations." Genome Res 13(3): 476-484.

Mack, C. P. and G. K. Owens (1999). "Regulation of smooth muscle alpha-actin expression in vivo is dependent on CArG elements within the 5' and first intron promoter regions." Circ Res 84(7): 852-861.

Mack, C. P., M. M. Thompson, et al. (2000). "Smooth muscle alpha-actin CArG elements coordinate formation of a smooth muscle cell-selective, serum response factor-containing activation complex." Circ Res 86(2): 221-232.

Murphy, K. C. (1998). "Use of bacteriophage lambda recombination functions to promote gene replacement in Escherichia coli." J Bacteriol 180(8): 2063-2071.

Myers, R. S. and F. W. Stahl (1994). "Chi and the RecBC D enzyme of Escherichia coli." Annu Rev Genet 28: 49-70.

Noll, S., G. Hampp, et al. (2009). "Site-directed mutagenesis of multi-copy-number plasmids: Red/ET recombination and unique restriction site elimination." Biotechniques 46(7): 527-533.

Poser, I., M. Sarov, et al. (2008). "BAC TransgeneOmics: a high-throughput method for exploration of protein function in mammals." Nat Methods 5(5): 409-415.

Warming, S., N. Costantino, et al. (2005). "Simple and highly efficient BAC recombineering using galK selection." Nucleic Acids Res 33(4): e36.

Wilson, C., H. J. Bellen, et al. (1990). "Position effects on eukaryotic gene expression." Annu Rev Cell Biol 6: 679-714.

Yeeles, J. T. and M. S. Dillingham (2010). "The processing of double-stranded DNA breaks for recombinational repair by helicase-nuclease complexes." DNA Repair (Amst) 9(3): 276-285.

Zhang, Y., F. Buchholz, et al. (1998). "A new logic for DNA engineering using recombination in Escherichia coli." Nat Genet 20(2): 123-128.

3

Defining the Deletion Size in Williams-Beuren Syndrome by Fluorescent *In Situ* Hybridization with Bacterial Artificial Chromosomes

Marc De Braekeleer[1,2,3] et al.*
[1]*Faculté de Médecine et des Sciences de la Santé, Université de Brest, Brest*
[2]*Institut National de la Santé et de la Recherche Médicale (INSERM), U613, Brest*
[3]*Service de Cytogénétique, Cytologie et Biologie de la Reproduction*
Hôpital Morvan, CHRU Brest, Brest
France

1. Introduction

Williams-Beuren syndrome (WBS, MIM No. 194050) is a contiguous gene deletion syndrome that was described independently by Williams et al. (1961) in patients with supravalvular aortic stenosis, growth retardation and an unusual facial appearance (Williams et al., 1961) and by Beuren et al. (1962) in patients having the same features as well as dental anomalies and friendly personality (Beuren et al., 1962).

The clinical picture of WBS includes a characteristic craniofacial dysmorphology with broad forehead, periorbital fullness, flat nasal bridge, broad nasal tip, long philtrum, full lips and lower cheeks, micrognathia, wide mouth and stellate irides, growth retardation, cardiovascular anomalies (mostly supravalvular aortic stenosis and pulmonary artery stenosis), mild to moderate mental retardation and unique behavioral and neurocognitive profile (relative preservation of linguistic abilities and gross visual-spatial processing dysfunction) (Morris et al., 1988; Pober, 2010). Its incidence is estimated at 1/7,500-1/20,000 (Grimm & Wesselhoeft, 1980; Morris et al., 1988; Stromme et al., 2002).

Williams-Beuren syndrome results from the hemizygous deletion of several genes, including *ELN* (elastin), encompassing 1.55 to 1.84 Mb on chromosome 7q11.23 (Ewart et al., 1993; Meng et al., 1998; Pober, 2010; Schubert & Laccone, 2006; Wang et al., 1999). WBS usually results from de novo deletion. Twenty-six to 28 genes have been identified within the WBS deletion region (Merla et al., 2010; Pober, 2010; Schubert, 2009).

---

*Audrey Basinko[1,2,3], Nathalie Douet-Guilbert [1,2,3], Séverine Audebert-Bellanger [4], Philippe Parent [4], Clémence Chabay-Vichot [1,4], Clément Bovo [1,2], Nadia Guéganic [1,2], Marie-Josée Le Bris [3], Frédéric Morel [1,2,3]
[1]*Faculté de Médecine et des Sciences de la Santé, Université de Brest, Brest, France*
[2]*Institut National de la Santé et de la Recherche Médicale (INSERM), U613, Brest, France*
[3]*Service de Cytogénétique, Cytologie et Biologie de la Reproduction, Hôpital Morvan, CHRU Brest, Brest, France*
[4]*Département de Pédiatrie et de Génétique Médicale, Hôpital Morvan, CHRU Brest, Brest, France*

The WBS deletion region is flanked by highly homologous clusters of genes and pseudogenes organized into low-copy-repeat (LCR) blocks. Unequal meiotic recombination during meiosis can lead to deletion of the WBS region. The unique genetic architecture of the region explains why the size of WBS deletion is almost the same in most of the patients (Baumer et al., 1998; Dutly & Schinzel, 1996; Pober, 2010; Schubert, 2009; Valero et al., 2000).

Several studies have investigated the size of the WBS deletion using multiplex PCR with several microsatellite markers (Brondum-Nielsen et al., 1997; Perez Jurado et al., 1996; Wang et al., 1999; Wu et al., 1998). Most of the patients were found to carry the 1.5 or 1.8 Mb deletion. In the present study, we performed fluorescent *in situ* hybridization (FISH) with Bacterial Artificial Chromosome (BAC) clones in an attempt to map the WBS deletion in 14 WBS patients.

2. Patients and methods

2.1 Patients

Fifteen patients were referred to the cytogenetic laboratory of the Brest University Hospital by medical geneticists, pediatricians and pediatric cardiologist for suspicion of WBS between 2002 and March 2011. The reasons for referral included cardiovascular anomalies, dysmorphism or a combination of signs suspecting WBS (Table 1). However, because of limited cell pellet available for one patient, the study could be performed only on 14 patients.

2.2 Conventional cytogenetics

Metaphase chromosomes were prepared from peripheral blood lymphocytes of the 14 patients after having obtained their parents' informed consent. Standard R banding chromosomal analyses were performed according to the standard procedures and the karyotypes described according to the International System for Cytogenetic Nomenclature (ISCN 2005).

2.3 FISH analyses with commercially available probe

A FISH study using the Vysis Williams Region Probe - LSI ELN SpectrumOrange/LSI D7S486, D7S522 SpectrumGreen was carried out on the metaphase preparations from all 14 patients, as recommended by the manufacturer (Abbott, Rungis, France).

The Williams Region Probe consists of a probe of approximately 180 kb in size for *ELN*, *LIMK1* and the D7S613 locus located in band 7q11.23, labeled in Spectrum Orange, and a control probe for the region containing loci D7S486 and D7S522 located in band 7q31, labeled in Spectrum Green.

2.4 FISH analyses with BAC clones

To delineate the extent of the deletion on chromosome 7, FISH analyses were carried out using BAC clones mapping to the long arm of chromosome 7 (7q11.23).

We identified the BAC clones of interest through the Human Genome Browser Database of the Genome Bioinformatics Group at the University of California at Santa Cruz (http://genome.ucsc.edu/). They were then ordered by Internet on the site of the Children's Hospital Oakland Research Institute in Oakland, California (http://bacpac.chori.org/).

Patient Nr.	1	2	3	4	5	6	7	8	9	10	11	12	13	14
Sex	F	M	F	M	F	M	M	F	M	M	F	M	F	F
Age at diagnosis	3m	2m	7m	1y4m	3m	1y7m	1y9m	9y11m	8m	2y9m	9m	2m	3m	2m
Reason for referral	Heart anomaly	Heart anomaly	Pulmonary artery stenosis	WBS suspicion	Aortic stenosis	Pulmonary artery stenosis	Heart anomaly	WBS suspicion	Dysmor-phism	Dysmor-phism	Heart anomaly	Dysmor-phism	Heart anomaly	Heart anomaly
Heart														
Ventricular septal defect	-	+	+	-	-	-	+	+	-	-	-	-	-	-
Supra valvular aortic stenosis	+	+	-	+	+	-	-	-	+	-	-	+	-	+
Pulmonary artery stenosis	-	+	+	-	+	+	-	-	-	-	+	+	+	+
Growth & Development														
IUGR	+	+	+	-	+	NA	-	-	+	+	-	-	-	-
Growth retardation	+	+	+	+	+	+	+	+	+	+	+	+	+	-
Psychomotor retardation	+	+	+	+	+	+	+	+	+	+	+	+	+	?
Digestive system														
Neonatal feeding difficulties	+	+	-	+	+	NA	+	-	+	+	+	+	+	-
Inguinal hernia	+	-	-	+	-	NA	-	-	-	+	-	+	-	-
Gastroesophageal reflux	+	-	-	+	+	NA	-	+	-	+	NA	+	+	+
Dysmorphology														
Periorbital oedema	-	+	-	+	+	+	+	+	-	-	+	+	-	+

Patient Nr.	1	2	3	4	5	6	7	8	9	10	11	12	13	14
Sex	F	M	F	M	F	M	M	F	M	M	F	M	F	F
Large mouth	-	-	+	+	-	-	+	+	+	+	+	+	-	+
Fine upper lip	-	+	-	-	+	-	+	+	+	+	+	+	+	-
Flat philtrum	-	+	-	-	-	-	-	+	+	+	+	+	+	+
Bulbous nose	-	+	+	+	+	+	+	+	+	+	-	-	-	+
Anteverted nares	-	-	-	-	+	-	+	+	-	-	+	-	+	+
Low-set dysplastic ears	-	+	-	-	-	-	-	-	+	+	-	-	+	-
Prominent forehead	-	+	-	+	-	-	-	-	-	-	+	-	+	-
ORL														
High arched palate or cleft palate	-	+	-	-	NA	NA	+	+	-	+	-	-	-	-
Ophthalmology														
Stellate irides	+	-	+	-	NA	NA	+	-	-	+	+	-	-	-
Strabismus ± astigmatism	NA	-	-	-	NA	NA	-	-	+	+	-	-	+	-

M: month; y: year
WBS suspicion: combination of signs (facial dysmorphology and cardiovascular anomaly) making the diagnosis highly probable
IUGR: intra uterine growth retardation
NA: not available
Patient # 14 - ?: patient still too young (4 months old)

Table 1. Clinical characteristics of the 14 patients with Williams-Beuren syndrome.

When received, bacterial cultures were prepared from a single colony picked from a selective plate in the presence of chloramphenicol. Plasmids were obtained from bacterial cultures grown in the presence of chloramphenicol (10 mg/L). After having lysed bacteria using SDS1%/NaOH 0.2 N, DNA was purified from RNA, proteins and other cellular contaminants. Probes were then labeled by nick translation in Spectrum Orange (Nick Translation Kit, Abbott, Rungis, France) or in FITC (Prime-it Fluor Fluorescence Labeling Kit, Stratagene, Amsterdam, Netherlands). All BAC clones were applied to normal lymphocyte metaphases to confirm their chromosomal location.

After hybridization, the slides were counterstained with 4-6-diamino-2-phenyl-indole-dihydrochloride (DAPI). The preparations were examined using a Zeiss Axio Plan Microscope (Zeiss, Le Pecq, France). Images acquisition was performed using a CCD camera and analyzed using the ISIS program (In Situ Imaging System) (MetaSystems, Altlussheim, Germany).

Twenty-two overlapping BACs covering the WBS deletion region and beyond (about 1.9 Mb) were applied on the 14 patients (Table 2).

BAC name	Centromeric start (Mb from telomere)	Telomeric start (Mb from telomere)	BAC length (bp)
RP11-48D17	72,557,570	72,705,204	147,635
RP11-483G21	72,688,731	72,875,946	187,216
RP11-614D7	72,737,634	72,926,932	189,299
RP11-101D2	72,814,548	72,991,974	177,427
RP11-598B14	72,850,682	73,042,752	192,071
RP11-622P13	73,002,073	73,181,256	179,184
RP11-73G23	73,035,513	73,184,825	149,313
RP11-148M21	73,153,473	73,327,919	174,447
RP11-1011-F11	73,166,799	73,342,468	175,670
RP11-1056I4	73,437,240	73,639,567	202,328
RP11-7M12	73,611,329	73,793,367	182,039
RP11-351B3	73,705,662	73,899,619	193,958
RP11-247L6	73,872,593	74,038,131	165,539
RP11-196F10	73,872,610	74,038,131	165,522
RP11-137E8	73,944,720	74,129,587	184,868
RP11-728M8	73,959,533	74,148,428	188,896
RP11-926D5	73,959,536	74,145,706	186,171
RP11-19F19	73,994,434	74,173,462	179,029
RP11-813J7	74,091,733	74,261,309	169,577
RP11-1105J19	74,128,975	74,321,491	192,517
RP11-1094P22	74,129,067	74,321,019	191,953
RP11-379L10	74,307,663	74,480,665	173,003

The base pairs position (bp) are predicted on Build 39 National Center for Biotechnology Information (http://www,ncbi,nlm,nih,gov) and assembly February 2009 by The UCSC Genome Browser Database (http://genome,ucsc,edu/index,html),

Table 2. BAC library used to delimitate the deletion size in the 14 WBS patients.

3. Results

Fourteen patients were included in the study (Table 1). There were 7 boys and 7 girls. The median age at the time of cytogenetic exam was 7.5 months. Except for a patient who was almost 10 years old, all the other patients were less than 3 years old at the time of cytogenetic analyses, the majority of them being less than 1 year old (9 of 14 patients).

Cardiovascular anomalies and facial dysmorphism were present in all patients, but for one without cardiovascular anomaly and another without dysmorphism (Table 1). Growth and psychomotor retardation was noted in all patients. Stellate irides and/or strabismus was also present in 7 of the 12 patients for whom the data was available. No cognitive nor personality profile was available for the patients, mainly due to their young age.

R-banding conventional cytogenetics showed a normal 46,XY or 46,XX karyotype whereas FISH using the Vysis Williams Region Probe revealed a deletion of the *ELN*, *LIMK1* and D7S613 locus in all 14 patients (Figure 1-A).

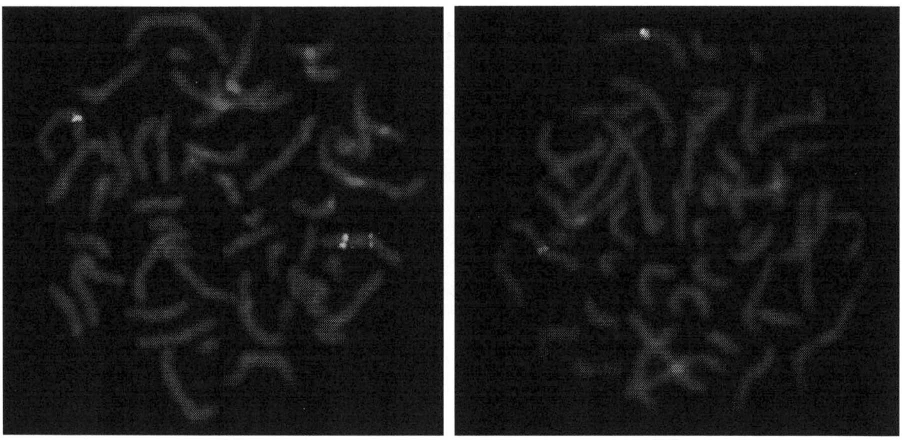

(a) (b)

Fig. 1. Example of FISH results in a patient with Williams-Beuren syndrome. (A) FISH using the commercially available probe (ELN in SpectrumOrange and D7S486, D7S522 in SpectrumGreen). (B) FISH using two BAC clones (RP11-614D7 in green and RP11-1105J19 in red).

Sequential FISH analyses with BAC clones were applied on metaphases of all 14 patients having a deletion of the Vysis Williams Region Probe (Figure 1-B). They showed the centromeric deletion boundary to be located in a 145 kb interval, between RP11-598B14, deleted in 13 patients (P1 to P13), and RP11-48D17, always present (Figure 2). The telomeric deletion boundary was found to be located in a 229 kb interval, between RP11-351B3, deleted in 13 patients (P1 to P13), and RP11-1105J19, always present. Therefore, the minimum and maximum estimated deletion sizes for these 13 patients (P1 to P13) were 1,048,937 bp and 1,423,771 bp, respectively. Using overlapping BAC clones, both centromeric and telomeric boundaries could be refined. FISH signals with BAC clones RP11-614D7 and RP11-101D2, located at the centromeric deletion boundary, and those RP11-

247L6 and RP11-137E8, located at the telomeric deletion boundary, showed decreased intensities. The centromeric and telomeric intervals were reduced to 36 kb and 93 kb, respectively, giving an estimated WBS deletion size of 1.17Mb.

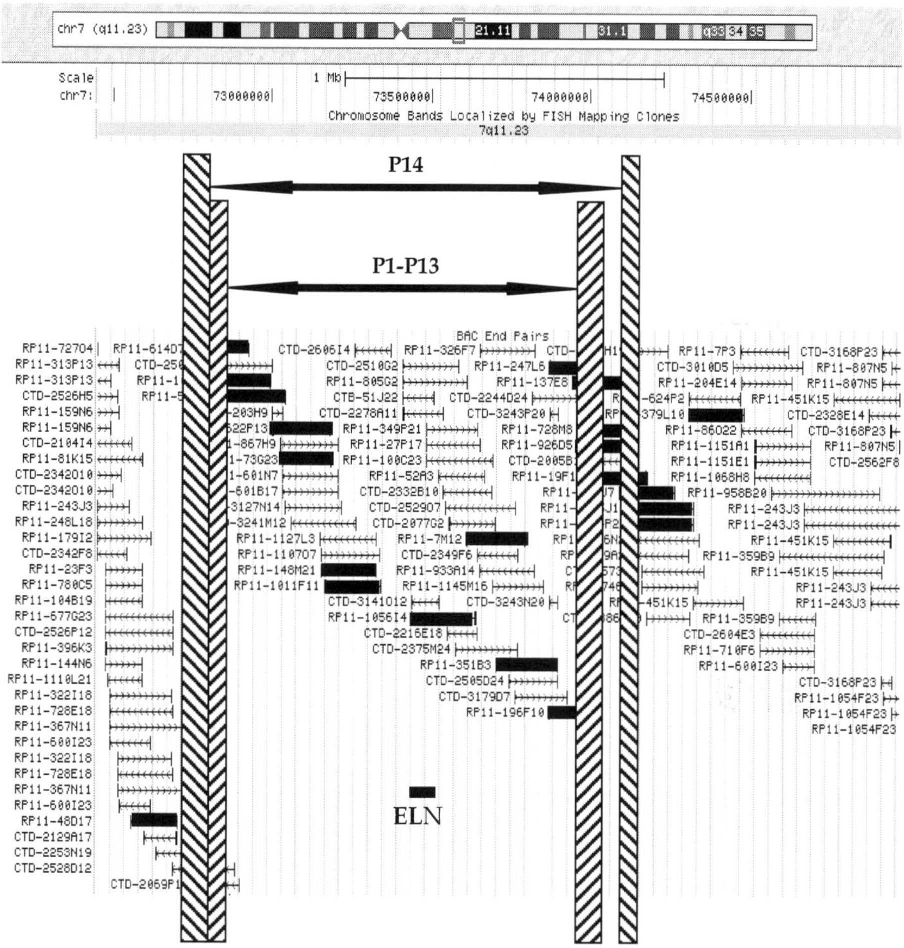

Fig. 2. Size of the deletions in the 14 WBS patients estimated by BAC clones (ELN: elastin gene)

The fourteenth patient (P14) was found to have a centromeric deletion boundary to be located in a 109 kb interval, between RP11-101D2 (deleted) and RP11-48D17 (present). Using overlapping BAC clones RP11-483G21 and RP11-614D7, the centromeric boundary was refined to a 77 kb interval. The telomeric deletion boundary was found to be located at the junction of two overlapping BAC clones (RP11-926D5 and RP11-1105J19). Therefore, the estimated WBS deletion size for this patient was 1.41Mb.

4. Discussion

Molecular genetic studies have shown that hemizygosity of the elastin (*ELN*) gene accounts for the cardiovascular abnormalities observed in WBS patients. The other signs (facial dysmorphism, growth and mental retardation, etc.) observed in WBS are likely to be due to hemizygosity of other genes flanking the *ELN* locus. Although the unique combination of signs and anomalies is highly evocative, WBS patients usually present phenotypic variability, which could be explained by different deletion size and, therefore, by different sets of genes showing hemizygosity.

Several studies tried to define the WBS critical deletion region. Using sequence tagged site (STS) markers flanking the *ELN* gene, Perez-Jurado et al. (1996) and Wu et al. (1998) looked for the minimal deletion region responsible for the WBS phenotype in 123 patients. Markers from D7S489B (also named D7S489U) to D7S1870 were shown to be consistently removed, defining a 2 cM deletion in all informative patients (Perez Jurado et al., 1996; Wu et al., 1998).

Using BAC and PAC (P1-derived artificial chromosome) clones, Meng et al. (1998) constructed a physical map of the common deletion region. They localized the centromeric and telomeric deletion breakpoints to two genomic clones (containing D7S489B and D7S1870) flanking a 1.5 Mb deletion interval (Meng et al., 1998).

Based on the work by Valero et al. (2000) who found that 3 large region-specific segmental duplication or low copy repeat (LCR) elements flanked the common deletion region (Valero et al., 2000), Bayes et al. (2003) defined two common deletion regions in a set of 74 patients with WBS. Most of the patients (95%) had a 1.55 Mb deletion whereas the remaining 5% exhibited a larger deletion of approximately 1.84 Mb (Bayes et al., 2003).

Botta et al. (1999) reported a deletion of about 850 kb in two patients showing the full spectrum of the WBS phenotype (Botta et al., 1999). Furthermore, the analysis of the deletion size in WBS patients with "incomplete" phenotype revealed even more heterogeneity (Antonell et al., 2010). Using quantitative real-time PCR to scan 2.5 Mb of the WBS deletion region at a resolution of 100-300 kb among 65 patients with strong clinical indication of having WBS, Schubert and Laccone (2006) found that 21 patients had a deletion in the WBS region. Nineteen patients had a deletion ranging from 1.4 to 1.8 Mb in size whereas one had a 200 kb deletion and the remaining one a 2.5 Mb deletion (Schubert & Laccone, 2006).

FISH using BAC clones has been extensively used to construct physical maps of the 7q region deleted in WBS and to define the commonly deleted region (Meng et al., 1998; Peoples et al., 2000; Perez Jurado et al., 1996; Wu et al., 1998). However, to our knowledge, no study has been conducted to determine the deletion size in WBS individuals using overlapping BACs. Indeed, the level of resolution is a limit to this technique. A BAC is considered as partially deleted by FISH analysis when the fluorescent signal of this clone is much stronger on one chromosome 7 than on the other. Furthermore, a BAC could be considered as present or absent (not partially deleted) if only a small part of the DNA sequence is removed or kept (usually less than 50 kb). Having resource to overlapping BAC clones can increase the level of confidence but some uncertainty will remain.

5. Conclusion

Thirteen patients had a deletion estimated at 1.17 Mb in size and the remaining patient a deletion size estimated at 1.41 Mb. All the patients had the "classical" WBS phenotype, but for the cognitive profile which was not evaluated. The margins of uncertainty of the centromeric and telomeric boundaries of the deletion were less than 100 kb in all 14 patients, a level obtained by microsatellite analysis or quantitative real-time PCR. Therefore, this approach is efficient to define the deletion size among WBS patients.

6. References

Antonell, A., Del Campo, M., Magano, L.F., Kaufmann, L., de la Iglesia, J.M., Gallastegui, F., Flores, R., Schweigmann, U., Fauth, C., Kotzot, D., & Perez-Jurado, L.A. (2010) Partial 7q11.23 deletions further implicate GTF2I and GTF2IRD1 as the main genes responsible for the Williams-Beuren syndrome neurocognitive profile. *Journal of Medical Genetics*, Vol. 47, pp. 312-320

Baumer, A., Dutly, F., Balmer, D., Riegel, M., Tukel, T., Krajewska-Walasek, M., & Schinzel, A.A. (1998) High level of unequal meiotic crossovers at the origin of the 22q11. 2 and 7q11.23 deletions. *Human Molecular Genetics*, Vol. 7, pp. 887-894

Bayes, M., Magano, L.F., Rivera, N., Flores, R., & Perez Jurado, L.A. (2003) Mutational mechanisms of Williams-Beuren syndrome deletions. *American Journal of Human Genetics*, Vol. 73, pp. 131-151

Beuren, A.J., Apitz, J., & Harmjanz, D. (1962) Supravalvular aortic stenosis in association with mental retardation and a certain facial appearance. *Circulation*, Vol. 26, pp. 1235-1240

Botta, A., Novelli, G., Mari, A., Novelli, A., Sabani, M., Korenberg, J., Osborne, L.R., Digilio, M.C., Giannotti, A., & Dallapiccola, B. (1999) Detection of an atypical 7q11.23 deletion in Williams syndrome patients which does not include the STX1A and FZD3 genes. *Journal of Medical Genetics*, Vol. 36, pp. 478-480

Brondum-Nielsen, K., Beck, B., Gyftodimou, J., Horlyk, H., Liljenberg, U., Petersen, M.B., Pedersen, W., Petersen, M.B., Sand, A., Skovby, F., Stafanger, G., Zetterqvist, P., & Tommerup, N. (1997) Investigation of deletions at 7q11.23 in 44 patients referred for Williams-Beuren syndrome, using FISH and four DNA polymorphisms. *Human Genetics*, Vol. 99, pp. 56-61

Dutly, F. & Schinzel, A. (1996) Unequal interchromosomal rearrangements may result in elastin gene deletions causing the Williams-Beuren syndrome. *Human Molecular Genetics*, Vol. 5, pp. 1893-1898

Ewart, A.K., Morris, C.A., Atkinson, D., Jin, W., Sternes, K., Spallone, P., Stock, A.D., Leppert, M., & Keating, M.T. (1993) Hemizygosity at the elastin locus in a developmental disorder, Williams syndrome. *Nature Genetics*, Vol. 5, pp. 11-16

Grimm, T. & Wesselhoeft, H. (1980) Zur Genetik des Williams-Beuren-Syndroms und der isolierten Form der supravalvulaeren Aortenstenose (Untersuchungen von 128 Familien). *Zeitschrift für Kardiologie*, Vol. 69, pp. 168-172

ISCN (2005). An International System for Human Cytogenetic Nomenclature. Basel: S. Karger.

Meng, X., Lu, X., Li, Z., Green, E.D., Massa, H., Trask, B.J., Morris, C.A., & Keating, M.T. (1998) Complete physical map of the common deletion region in Williams syndrome and identification and characterization of three novel genes. *Human Genetics*, Vol. 103, pp. 590-599

Merla, G., Brunetti-Pierri, N., Micale, L., & Fusco, C. (2010) Copy number variants at Williams-Beuren syndrome 7q11.23 region. *Human Genetics*, Vol. 128, pp. 3-26

Morris, C.A., Demsey, S.A., Leonard, C.O., Dilts, C., & Blackburn, B.L. (1988) Natural history of Williams syndrome: physical characteristics. *Journal of Pediatrics*, Vol. 113, pp. 318-326

Peoples, R., Franke, Y., Wang, Y.K., Perez-Jurado, L., Paperna, T., Cisco, M., & Francke, U. (2000) A physical map, including a BAC/PAC clone contig, of the Williams-Beuren syndrome--deletion region at 7q11.23. *American Journal of Human Genetics*, Vol. 66, pp. 47-68

Perez Jurado, L.A., Peoples, R., Kaplan, P., Hmal, B.C.J., & Francke, U. (1996) Molecular definition of the chromosome 7 deletion in Williams syndrome and parent-of-origin effects on growth. *American Journal of Human Genetics*, Vol. 59, pp. 781-792

Pober, B.R. (2010) Williams-Beuren syndrome. *New England Journal of Medicine*, Vol. 362, pp. 239-252

Schubert, C. (2009) The genomic basis of the Williams-Beuren syndrome. *Cellular and Molecular Life Sciences*, Vol. 66, pp. 1178-1197

Schubert, C. & Laccone, F. (2006) Williams-Beuren syndrome: determination of deletion size using quantitative real-time PCR. *International Journal of Molecular Medicine*, Vol. 18, pp. 799-806

Stromme, P., Bjornstad, P.G., & Ramstad, K. (2002) Prevalence estimation of Williams syndrome. *Journal of Child Neurology*, Vol. 17, pp. 269-271

Valero, M.C., de Luis, O., Cruces, J., & Perez Jurado, L.A. (2000) Fine-scale comparative mapping of the human 7q11.23 region and the orthologous region on mouse chromosome 5G: the low-copy repeats that flank the Williams-Beuren syndrome deletion arose at breakpoint sites of an evolutionary inversion(s). *Genomics*, Vol. 69, pp. 1-13

Wang, M.S., Schinzel, A., Kotzot, D., Balmer, D., Casey, R., Chodirker, B.N., Gyftodimou, J., Petersen, M.B., Lopez-Rangel, E., & Robinson, W.P. (1999) Molecular and clinical correlation study of Williams-Beuren syndrome: No evidence of molecular factors in the deletion region or imprinting affecting clinical outcome. *American Journal of Medical Genetics*, Vol. 86, pp. 34-43

Williams, J.C., Barratt-Boyes, B.G., & Lowe, J.B. (1961) Supravalvular aortic stenosis. *Circulation*, Vol. 24, pp. 1311-1318

Wu, Y.Q., Sutton, V.R., Nickerson, E., Lupski, J.R., Potocki, L., Korenberg, J.R., Greenberg, F., Tassabehji, M., & Shaffer, L.G. (1998) Delineation of the common critical region in Williams syndrome and clinical correlation of growth, heart defects, ethnicity, and parental origin. *American Journal of Medical Genetics*, Vol. 78, pp. 82-89

4

Functionalizing Bacterial Artificial Chromosomes with Transposons to Explore Gene Regulation

Hope M. Wolf[1,2], Oladoyin Iranloye[1,3],
Derek C. Norford[1] and Pradeep K. Chatterjee[1,3]
[1]*Julius L. Chambers Biomedical/ Biotechnology Research Institute*
[2]*Department of Chemistry, University of North Carolina at Chapel-Hill, Chapel-Hill, NC*
[3]*Department of Chemistry, North Carolina Central University, Durham*
USA

1. Introduction

Strategies for altering sequences in large DNA inserts in BACs are fundamentally different from those traditionally used for small plasmids. Two factors are primarily responsible for this: the existence of a high multiplicity of sites in BACs recognized by DNA modifying enzymes, such as restriction endo-nucleases, as well as the brittle nature of large duplex DNA that is not packaged with DNA-binding proteins in the test-tube. The large number of DNA fragments generated by the common restriction enzymes, and the unavailability of robust separation techniques to isolate and keep track of the relative order of the pieces; precludes using the "cut-and-paste" mechanism to alter DNA sequences in BACs. Instead DNA recombination *in vivo* has become the method of choice for modifying BACs. Although the net result again is cutting and re-joining of DNA, the entire process is concerted in the bacterial host with no free ends of DNA to go astray; and occurs primarily in a nucleoprotein complex that protects it from shear forces which otherwise would break the large DNA during manipulations *in vitro.*

Two separate approaches based upon using sequence homology for recombining DNA were developed independently to alter sequences in BACs. The first method introduces the major recombination function of *E.coli,* RecA, back into the severely recombination deficient host DH10B, originally engineered to propagate vertebrate DNA in BACs (1). The second approach introduces recombination functions of phage λ, namely red α, red β and red γ into the DH10B host, and utilizes recombination of homologous sequences of shorter length to insert exogenous DNA cassettes into BACs (2). Both methods have been widely used to engineer a variety of alterations in BAC DNA, such as introducing reporter gene cassettes into the genomic insert, mutating sequences at a target site, and introducing loxP sites (1-11).

A different strategy for modifying BAC DNA also uses recombination, but does not require targeting vectors to carry sequences homologous to those in the genomic insert to introduce exogenous DNA cassettes. Insertions of the bacterial Tn10 transposon can introduce exogenous DNA, including lox sites, at random locations in the BAC (12, 13). Site-specific recombination using the Cre-lox system on the other hand, can deliver reporter genes and other exogenous DNA cassettes such as sequencing primer sites, mammalian cell-selectable antibiotic resistance genes, enhancer-traps and sequences specific to the vertebrate transposon Tol2 precisely at the ends of the genomic DNA insert in BACs (14-16). It is significant that the recombinases involved in either of these approaches, Tn10-transposase and Cre protein respectively, do not act upon sequence repeats and/or other recombinogenic sites in the genomic DNA insert to rearrange it. This particular characteristic should make the approach applicable to a wider variety of BACs in the public domain, including those containing repetitive sequences [see reference (16) for an example].

Insertions of the Tn10 transposon into DNA of BACs from a wide variety of vertebrate genome libraries appear to be random, demonstrating little sequence specificity for transposition (16). It is unclear whether this lack of sequence specificity arises from the absence of selective pressure evolutionarily for insertions of a prokaryotic transposon into vertebrate DNA, because insertions into prokaryotic DNA have long been known to prefer a somewhat degenerate nevertheless consensus insertion site (17). The minor sequence preferences for insertions of Tn10 observed in BACs probably have more to do with availability of sites for Tn10 insertions in HU protein-packaged vertebrate DNA in the bacterium than specificity for sequences. On occasion, incorporation of Tn10 into insert DNA of a rare BAC clone displays apparent sequence selectivity (18). However, this was clearly shown to be due to a clonal selection process that arose from picking single colonies of transposon plasmid transformed BAC clones which had induced the transposase gene prior to actual induction with Isopropyl β-D-1-thiogalactopyranoside (IPTG). Slight modification of the procedure, that has since recommended inducing a large pool of transposon plasmid transformed BAC colonies instead of a single one, has rectified this potential problem completely (18). A detailed description of the transposon retrofitting approach for BACs follows.

1.1 Methodology for Tn10 transposon retrofitting of BACs

The Tn10 transposon modification procedure for BACs is conceptually simple, as illustrated in Figure 1. A loxP sequence is placed within the 70 base inverted repeat ends of the bacterial Tn10 mini-transposon (shown in the lower panel of Figure 1).

The plasmid DNA carrying the transposon is introduced into the same cell that houses the BAC plasmid and the transposase gene, located outside the inverted repeat ends in the transposon plasmid, is induced. Upon induction the transposase protein excises the DNA cassette flanked by the inverted repeats (shown as green and pink boxes L & R in the transposon plasmid) and inserts it into other nearby DNA in the bacterial host, including the bacterial genome and the BAC DNA. Insertions of the Tn10 into BAC DNA occur in either orientation, and are irreversible because the transposase gene is left behind in the excised transposon plasmid and destroyed. Important considerations that dictate subsequent steps in the procedure include 1) damage to the host genome from Tn10 insertions, and 2) the fraction of BACs actually modified is relatively low ~ 1 in 10,000. Therefore efficient steps are required to recover the low percentage of BACs containing insertions and subsequently transfer these into a new host. Both these challenges are met by packaging the Tn10-modified BAC DNA in phage P1 heads. Therefore the cells containing the BAC DNA with transposon insertions are infected with P1 phage after induction of the transposase gene with IPTG [see reference (19) for a detailed description of the procedure].

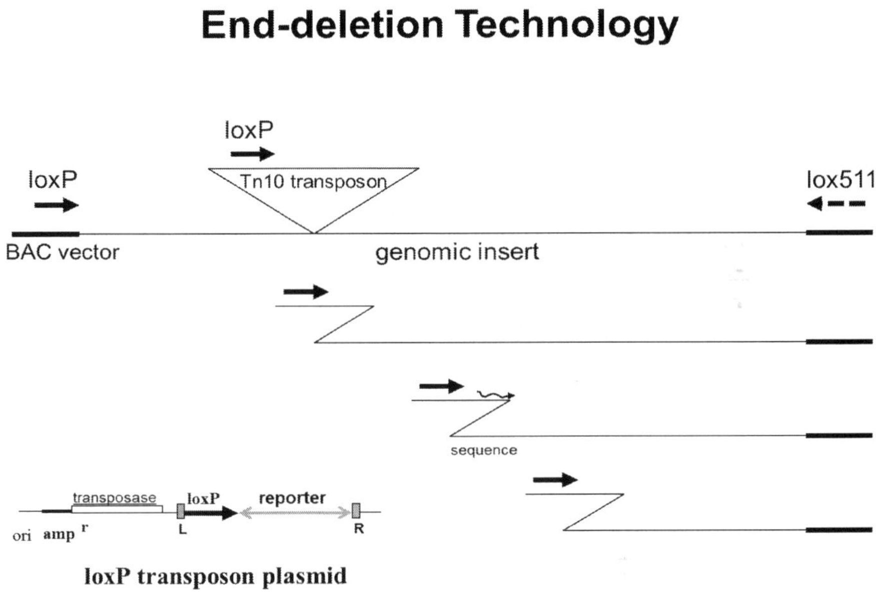

Fig. 1. Legend: Schematic representation of the BAC end-deletion technology using a loxP transposon. The inverted triangle represents the loxP transposon, which is shown in greater detail in the lower panel. The 70 bp inverted repeat ends of the transposon are indicated by the green and pink boxes marked L and R respectively. The thick black arrow represents the loxP site in both the transposon and BAC vector DNA. The broken arrow represents the lox511 site in BAC vector. Because transposon insertions are rare, single random transpositions into the BAC DNA occur when the transposase gene is induced. Upon Cre recombination, the transposon-inserted loxP of identical orientation to the loxP endogenous to the BAC generates a deletion of the DNA between them. Inversions are not shown. The pool of BAC DNA molecules therefore generates a library of end-deleted BACs from the random insertions of a loxP transposon into BAC DNA molecules.

Infection with P1 phage serves an additional purpose. The phage expresses Cre protein early in its life cycle to circularize the otherwise linear DNA within the phage head. Newly synthesized Cre protein acts *in trans* to also recombine the loxP site transposed into the BAC genomic insert with the loxP endogenous to the BAC and located at one end of the insert DNA [see step 1, Figure 2]. Thus the transposed loxP site of one orientation produces a deletion from one end of the genomic insert, while the loxP inserted in the opposite orientation simply inverts the DNA between it and the one endogenous to the BAC. Because BACs in all modern libraries carry insert DNA of average size ~ 160 kb, and because the amount of DNA that can be packaged in a P1 head is ~110 kb (20), the P1 headful packaging step can serve as a selection strategy to isolate transpositions of loxP in the orientation identical to the endogenous one (12), as only that allows reducing the BAC DNA length to less than ~110 kb. Only those end-deletions that reduce the length of BAC DNA to less than ~110 kb are rescued, as shown in Figure 2.

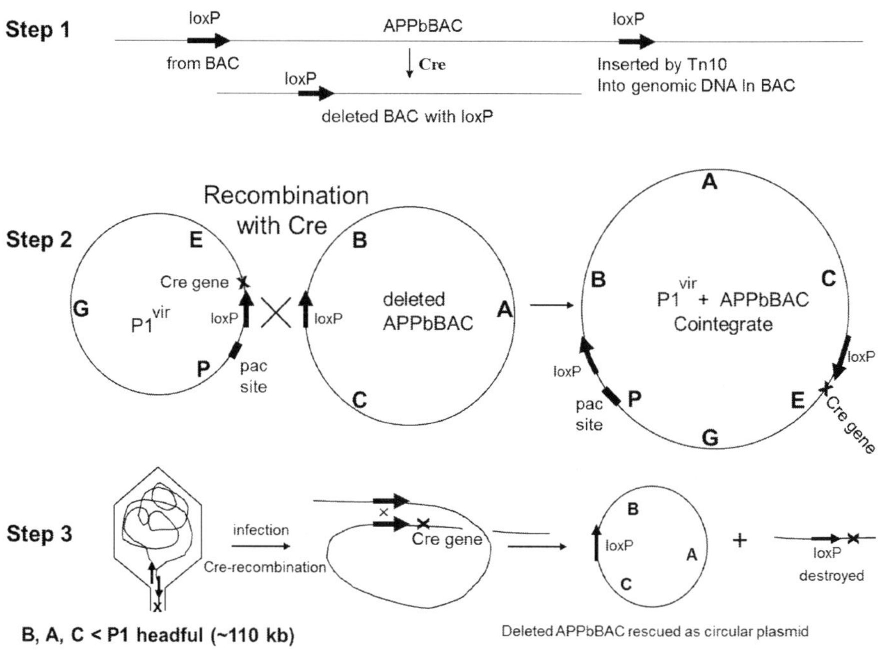

Fig. 2. Legend: Schematic representation of the DNA recombinations occurring in the transposon retrofitting of BACs. Step 1 shows creation of a deletion in the BAC DNA by Cre recombination of the transposed and endogenous loxP sites. Step 2 shows the Cre recombination of circularized phage P1 DNA (P, G, E) with the end-deleted BAC DNA (B, A, C) to generate the co-integrate. The phage packaging site "pac site" and the Cre gene are indicated in the phage DNA by the thick solid bar and the X, respectively. Step 3 shows the BAC DNA packaged in the phage P1 head, its recombination after entry into cells by newly synthesized Cre protein, and circularization into a BAC plasmid. If the length of DNA (B, A, C) exceeds the headful packaging capacity of ~110 kb, then the second loxP site (indicated by the thick arrows) would not fit in the phage head and the DNA cannot circularize by loxP-Cre recombination after entry into cells. This leads to the BAC DNA not being rescued. Note that the Cre gene is also lost upon circularization of the truncated BAC DNA.

Packaging of the end-deleted BAC DNA occurs from a large co-integrate plasmid in the cell which is described in detail elsewhere (21, 22). In co-integrate formation, the P1 phage DNA effectively contributes a packaging location named the "pac site" which is recognized by the phage packaging proteins [shown in step 2, Figure 2]. The co-integrate DNA is thought to replicate by a "rolling circle" mechanism. A cut is made in the newly synthesized co-integrate DNA at the "pac site", and the DNA end corresponding to piece loxP-B, A, C is stuffed into newly assembled empty heads of phage P1. Packaging of DNA from co-integrate is directional and the second cleavage is made after the P1 head is full regardless of sequence using a "headful-cleavage" mechanism (21, 22). Note that if Cre recombination of the transposed loxP with the loxP endogenous to the BAC did not reduce the length of BAC DNA segment B, A, C, to less than ~110 kb, this piece of BAC DNA inside the phage head would not be flanked by loxP on both ends [see step 3 of Figure 2]. In the absence of loxP sites flanking BAC DNA on both ends, the DNA would be unable to circularize by Cre recombination and would be destroyed when introduced into cells. Insertions of loxP in the other orientation, that cause inversion of the DNA, do not reduce the size of segment B, A, C to less than 110 kb. Consequently, the BAC DNA in P1 heads from inversions is destroyed upon entry into cells.

The bacterial lysate containing P1 heads packaged with end-deleted BAC DNA is treated with chloroform. This treatment not only facilitates lysis of P1 infected recombination deficient DH10B but also kills all cells harboring intact transposon and BAC plasmids. This killing is important for the selection of transposon inserted BACs in the next stage when antibiotic selection is employed, as otherwise cells without transposon insertions into BACs but merely carrying the two intact plasmids would get selected. The lysate after chloroform treatment is used to infect fresh bacteria. Upon entry of the phage packaged BAC DNA into cells; Cre protein is expressed from the DNA end adjacent to the second loxP site [indicated as X in phage head and co-integrate], but only transiently, because the Cre gene is lost after Cre-loxP recombination to circularize the linear BAC DNA [Step 3, Figure 2]. Cells are plated on LB plates containing chloramphenicol to select for clones of end-deleted BAC. Note that the phage DNA segment [P, G, E] from the co-integrate is incompletely packaged, if at all, and is destroyed after the loxP-Cre circularization in the new host. Even the rare phage particle containing a complete phage genome is unlikely to replicate or survive because of the chloramphenicol selection on the plates.

Selecting with a single antibiotic for the BAC DNA is sufficient for the first round of end-deletions, using either a loxP or a lox511 transposon. Additional selection for transposition of lox sites is not necessary in this first round because the P1 headful packaging itself serves as the selection for the low frequency of lox-site transpositions in the previous step [see references (23, 24) for detailed discussions].

1.2 Analyzing retrofitted/ end-deleted BACs

BACs deleted from an end, or inserted with exogenous DNA cassettes at the newly created ends of genomic inserts, are analyzed by isolating their DNA and separating them by Field Inversion Gel Electrophoresis (FIGE). The DNA in BAC clones is easily isolated by simple mini-prep procedures, digested with Not I enzyme and analyzed on FIGE (13, 25), as shown in Figure 3. High throughput formats for preparing DNA from BAC deletion clones in 96 well chambers, suitable for subsequent FIGE analyses and end sequencing, have also been described earlier (13, 25).

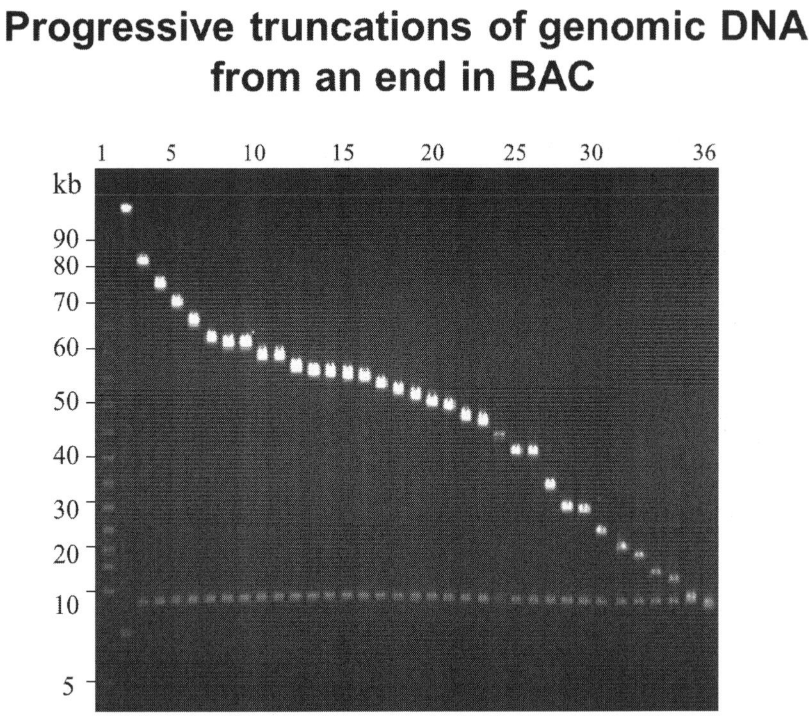

Fig. 3. Legend: FIGE display of an array of BAC deletion clone DNAs. DNA was isolated by the miniprep procedure from clones of a library of end-deleted BACs generated using a loxP transposon. The DNA was digested with Not I enzyme before electrophoresis. Lane 1 shows a 5 kb ladder, and lane 2 shows the DNA from the starting BAC clone.

1.3 Scoring authentic lox-Cre recombinations by unique size of BAC vector DNA upon Not I digestion

Whether the BAC DNA of reduced size arose from an authentic lox-Cre recombination, as opposed to an internal deletion between sequence repeats or other recombinogenic sites in the genomic insert, is one of the early determinations that need to be made. Note that lox-Cre independent deletions can also be packaged by the P1 headful mechanism when there is no selection for lox site transposition *per se;* such deletions are also efficiently isolated (23). Therefore all our loxP and lox511 transposons have been designed to alter the size of the Not I-Not I BAC-vector DNA band that arises from a Not I enzymatic digest of the deletion clone DNA [see references (23, 24, 27) for discussion]. An end-deletion from Cre-loxP recombination alters the size of BAC vector DNA, whereas deletions arising from within the genomic insert do not. End-deletions made with our lox511 transposons remove the Not I site at the lox511 end in the original BAC, resulting in a linear BAC with no separate vector DNA band (24).

Once authenticity of the retrofitted/ end-deleted BAC clone is established, the exact location of insertion of the transposon on genomic DNA is determined by sequencing the BAC DNA directly with unique primer sites located on either the loxP or lox511 transposon ends that remain after deletion formation (13, 24).

1.4 Truncations from the other end of insert DNA in BACs

The genomic DNA insert in BACs is flanked by a loxP site at one end and a mutant lox511 site at the other (24, 26). Modern genome libraries of DNA from numerous organisms such as the mouse, rat, human and zebrafish have used the vector pBACe3.6 or its derivatives [reference (26), link for pTARBAC2.1 vector *http://bacpac.chori.org/ptarbac21.htm*], and all share this characteristic (27-31). End-deletions have been made specifically from the lox511 side of genomic insert by a similar approach using a lox511 Tn10 mini-transposon (24).

It is important to note that second round deletions from the opposite end of insert DNA requires selecting for the transposition event in addition to the BAC DNA: because first round deletions are selected by P1 headfull packaging, all starting BACs for second round deletions would be less than ~110 kb, and hence require selecting for the low frequency of transposition [see reference (23, 24) for detailed discussions].

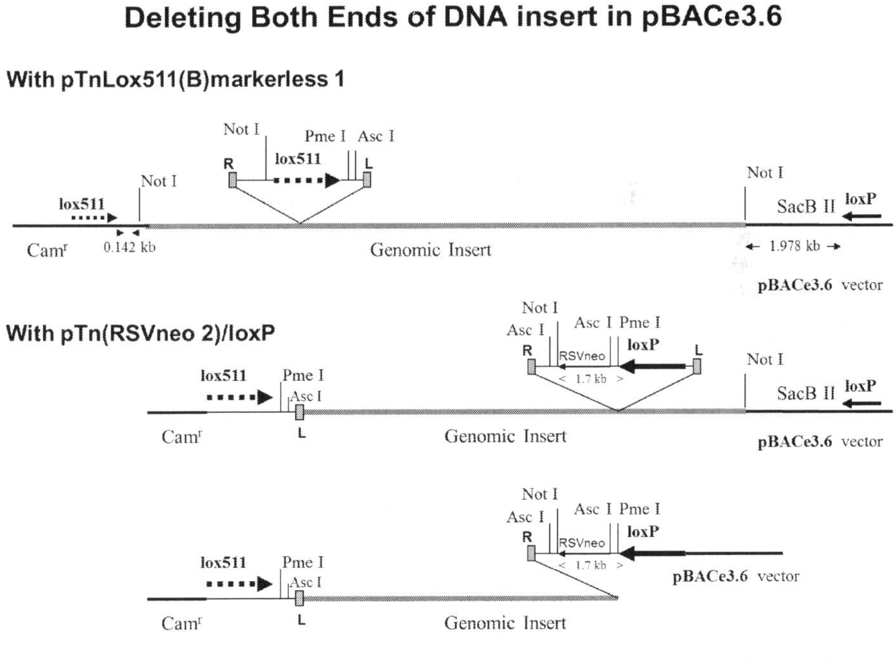

Fig. 4. Legend: Schematic representation of end-deletions generated from both ends of DNA insert in a BAC clone. The upper panel shows truncations from the lox511 end using a lox511 transposon. The largest clone from that library was then truncated from the loxP end with a loxP transposon. Figure is adapted from reference (24).

1.5 Potential Cre-mediated cross-recombination of loxP & lox511 sites does not occur in our method

The 34 bp sequence in the mutant lox site, lox511, differs by one nucleotide in the spacer region from that of loxP [see references (24, 32), for actual sequence of lox511]. Varying degrees of promiscuity in recombining different mutant *lox* sites, including the lox511 mutant, with wild type loxP have been reported in previous studies; using both partially purified Cre-extracts *in vitro* (32-36), or Cre over-expressed in cells (36-38). For example, cross recombination between loxP and lox511 has been reported to occur at efficiencies ranging from 5 to 100 % under those experimental conditions that express Cre constitutively (33, 35-38). We have not observed such cross-recombination, and high levels of stringency *in vivo* in recombining identical *lox* sites (21, 24), or *lox* sites with at least identical spacers (39), have been achieved with Cre protein expressed from its native source namely, a phage P1 infection (21, 24, 39). Depending on whether a loxP or a lox511 transposon is used, truncations from the corresponding lox- end can be made readily and exclusively with high stringency [shown schematically in Figure 4]. Thus truncations of genomic DNA from either end are not only efficient, but are highly specific for that end.

1.6 Exogenous DNA cassettes can be introduced precisely at one or both newly created BAC ends

An important feature of the end-deletion technology is the ease of determining exactly where in the BAC the loxP-transposon had inserted to create the truncation: primers for sequencing the new BAC end have been designed into the loxP and lox511 transposon-ends remaining after the lox-Cre recombinations [see Figure 1 and references (13, 24)], and these have been used to generate ~600 base reads by sequencing the retrofitted BAC DNA directly. An additional feature is the ability to introduce reporter genes and other DNA cassettes precisely at the new end created in the BAC clone. The sequence in front of the loxP or lox511 arrows, as shown in Figures 1 and 4, is retained after the recombination event that creates the deletion [note that orientation of arrow refers to directionality of loxP and lox 511 sequences]. This particular feature has been utilized to place numerous DNA cassettes, such as: i) mammalian cell-selectable antibiotic resistance genes in the BAC (24), ii) a basal promoter-containing EGFP gene in the BAC to serve as an enhancer-trap to functionally localize potential gene-regulatory elements further upstream (14), and iii) introduce iTol2 ends to generate functionalized BAC DNA ready for integration into zebrafish or mouse chromosomes (15).

The ability to truncate either end of a BAC insert progressively while keeping the other end intact is useful to a variety of mapping experiments. Thus genetic markers and polymorphisms have been mapped on a physical map of the chromosome using this technology (13, 16). Long-range gene regulatory elements have also been mapped functionally using the end-deletion approach by either generating transgenic mice with EGFP functionalized BAC DNA (40), or electroporating patient-derived cells in culture (41). The Nkx2-5 gene containing BAC from a mouse library was functionalized with EGFP reporter gene using the phage λ red-recombination system, and then truncated progressively from the far upstream 5′ end of the Nkx2-5 gene using a loxP transposon to identify transcription-enhancing factor binding sites (40). Monitoring mRNA levels of the gene of interest allows circumventing the use of a reporter gene such as EGFP in the BAC or PAC DNA, and a series of upstream deletions in a PAC clone was used to functionally identify potential far-upstream binding sites for transcription factors in the human Hox11 gene (41).

1.7 Simultaneous insertion of reporter gene & truncation of DNA from BAC ends

The two separate steps of introducing the EGFP reporter gene and truncating far upstream sequences of the gene of interest in the BAC can be done simultaneously using lox-Tn10 transposons (14). A basal promoter-containing EGFP gene was designed to serve as an enhancer-trap, and placed in front of the loxP arrowhead such that the cassette was retained after the end-deletion of the BAC DNA [see upper panel of Figure 5 for enhancer-trap transposon]. Progressive truncations from one end of the genomic insert with this enhancer-trap transposon generated a library of BAC DNA molecules that differ in the position of the enhancer-trap cassette with respect to the start site of transcription of the gene. The collection of well characterized BAC DNA can be introduced individually into zebrafish embryos for expression of the gene in specific tissues (14).

Such enhancer-trap containing BACs can be re-fitted again at the opposite end with DNA cassettes carrying repeat ends of the vertebrate transposon Tol2 as described below.

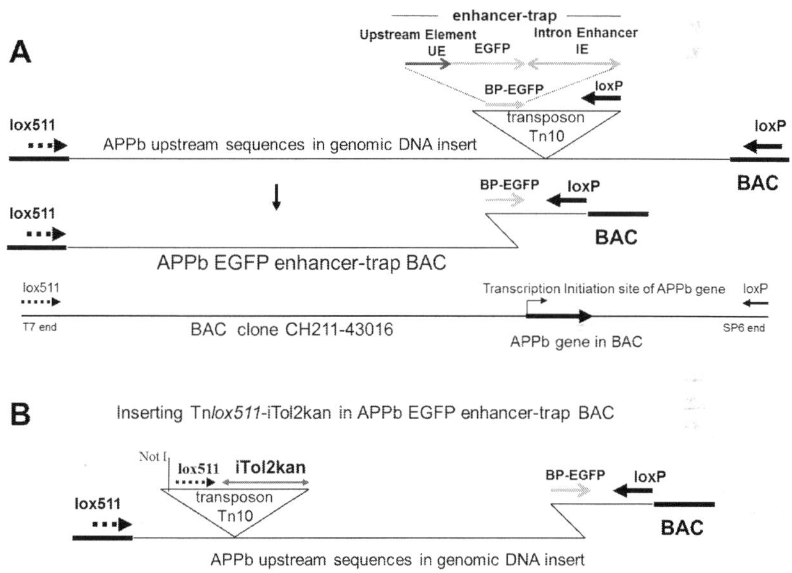

Fig. 5. Legend: Schematic representation of deletion formation in starting BAC APPb EGFP enhancer-trap by pTn*lox511*-iTol2kan: Panel A shows generation of starting BAC APPb EGFP enhancer-trap by insertion of the enhancer-trap transposon (inverted triangle), and the location of APPb gene within the BAC clone. Panel B: Shown schematically is the insertion of the Tn*lox511*-iTol2kan transposon into the enhancer-trap APPb BAC DNA. The thick black arrows represent loxP and lox511 sites (discontinuous arrow). DNA cassettes in front of the loxP or lox511 arrows are retained after the loxP-loxP or lox511-lox511 recombinations by Cre. The enhancer-trap DNA cassette is positioned in front of the loxP arrow (panel A), while the iTol2 cassette is located in front of the lox511 arrow. The colored arrowheads pointing outward in the iTol2 cassette correspond to the 200 bp inverted repeat end R and 150 bp inverted repeat end L of the Tol2 transposon [see Suster et al 2009 ref (42)]. The kanamycin resistance gene is between the Tol2 inverted repeat ends.

1.8 BAC transgenesis in zebrafish and mice using Tol2

BACs have been integrated into the germlines of zebrafish and mice using the vertebrate transposon Tol2 system (42). Tol2 ends, in the inverted orientation and flanking a 1 kb spacer DNA (iTol2), were introduced into the BAC DNA within the bacterial host using recombination of homologous sequences between the targeting vector and the genomic insert. This approach used to introduce the iTol2 cassette can have unintended consequences: other sequence repeats existing in BACs from vertebrate DNA libraries were likely to rearrange as well. This would complicate the introduction of iTol2 cassettes at best, and were likely to present major challenges in BACs spanning chromosomal loci that contained highly repetitive DNA, such as the Npr3 gene locus analyzed earlier (16). Therefore a simpler and more flexible system was developed using our Tn10 transposon approach to deliver iTol2 ends into BACs. These iTol2-end containing BACs are suitable for transgenesis into zebrafish or mouse embryos, and have recently been reported (15).

The iTol2 DNA cassette was placed in front of a loxP or lox511-Tn10 transposon (15). Progressive truncations from an end of the genomic insert with this iTol2-Tn10 transposon generated a library of BAC DNA molecules that differed in the position of the iTol2 cassette with respect to upstream regulatory elements of the Amyloid Precursor Protein (APPb) gene in the BAC. The collection of well characterized BAC DNA was introduced into zebrafish embryos individually for expression of EGFP in neurons (Figure 6).

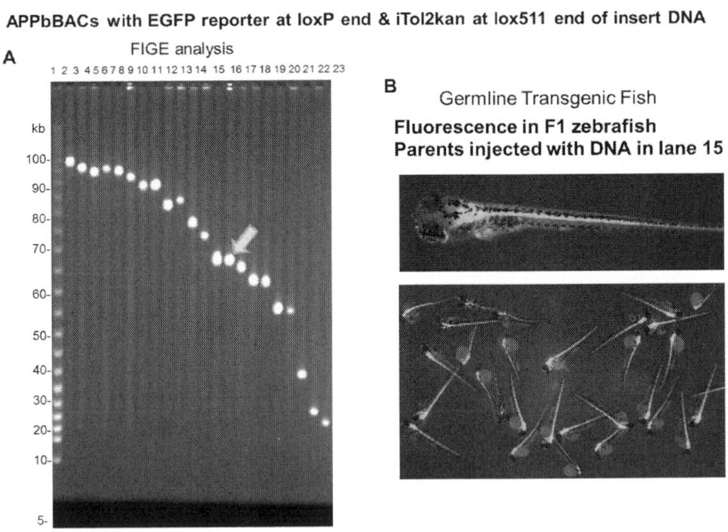

Fig. 6. Legend: Panel A: FIGE analysis of an array of APPb BACs with EGFP containing Enhancer-trap at loxP end and iTol2kan cassette at lox511 end of DNA insert. The BAC DNA was digested with Not I before FIGE. Note that there is no vector DNA band because the lox511-lox511 deletion with Cre removes both the Not I site in front of the BAC lox511 and the Not I site behind the lox511 in the transposon shown in panel B of Figure 5. Panel B: EGFP fluorescence in Germline transgenic F1 zebrafish obtained from injecting eggs (F0 zebrafish) with the BAC DNA shown by the blue arrowhead in lane 15.

2. Exploring gene regulation by distal *cis*-acting sequences using BACs

2.1 Rationale for using BACs to understand gene regulation

BACs offer a large span of DNA that can house most, if not all, the sequences recognized by DNA-binding regulatory proteins that act in concert to regulate expression of the gene. But more importantly, the binding sites of regulatory proteins in the BAC DNA exist in their chromosomal contexts, with sequences flanking *cis*-acting sites that have evolved for millions of years, preserved. This cannot be said of small plasmid constructs used for expressing genes where multiple, easily-recognizable distal *cis*-acting sequences are excised from their surrounding DNA in the chromosome and cobbled together to create artificial junctions.

2.2 Why is this important in regulation of gene expression?

We know that both the rate of a reaction, as well as the equilibrium that a reversible reaction establishes, are related inversely and logarithmically to the activation energy or the free energy difference between the reactants and products, respectively. It implies that small changes in the binding affinities of DNA-binding regulatory proteins, important for regulation of gene expression, can profoundly affect the rates and equilibrium established in the cell. Also, gene regulatory proteins in the cell do not exist as monomeric units, but as components of multiple different protein complexes with different physical dimensions destined for a variety of purposes in the cell. Thus it can be argued that the same protein in a multimeric complex might not be able to play identical gene-regulatory roles if the sequences flanking its binding site are altered compared to when they exist in their chromosomal context. It is conceivable that some of this disparity in DNA-binding affinities resulting from binding to sites with altered flanking sequences could eventually manifest itself in altered tissue specificity of expression of the gene.

There is also the issue of how one goes about selecting potential gene-regulatory elements to stitch together in a small plasmid designed for expressing a gene of interest. While one can make informed guesses one cannot be unbiased, as much of the players involved in the fine-tuning of gene expression remain largely unknown. Using BACs functionalized with suitable reporter genes, and integrated into the chromosome, to explore regulation of genes by *cis*-acting elements appears to circumvent many of these difficulties.

We have developed one such approach using BACs retrofitted with enhancer-traps (14). The overall effect of using this enhancer-trap technology in BACs is somewhat akin to scanning the BAC DNA from one end to the other with a "mine-sweeper" that is capable of unearthing transcription-enhancing factor binding sites. Functional comparisons between individual enhancer-trap BACs are thus more meaningful as the modifications in each remain constant. It is important to note that this approach to identifying gene-regulatory elements is likely to be less biased than a targeted approach, because there is no prior selection of sequences for testing their gene-regulatory potential (14, 15). In situations where regulatory function is conserved without sequence similarity (43), choosing the correct sequence to test might be a hit-or-miss phenomenon with a targeted approach. Thus using enhancer-traps to scan BACs offers an effective and relatively unbiased alternative to other targeted approaches to functionally map *cis*-acting gene regulatory elements. Important transcription enhancing elements have been discovered using this approach in non-coding DNA from the intron, and 30 kb upstream of the APPb gene transcription start site [reference (14), and Shakes, Du, Sen, Abe, Kawakami, Wolf, Hatcher, Norford and Chatterjee manuscript in preparation].

The BAC enhancer-trap technology described above has three additional features that should prove beneficial: i) allows sampling much larger sequences of DNA, and consequently multiple discontinuous regulatory domains simultaneously compared to small plasmids, ii) the context of regulatory DNA with respect to the gene and chromosome is preserved and, iii) can be used with BACs in established libraries from a wide variety of organisms, and tested in several species. Although the methodology does not allow generation of internal deletions, truncations from the end opposite to the enhancer-trap can be made with a lox511 transposon [as seen when iTol2-end DNA cassettes were introduced at the other end of BAC DNA, (Figures 5 and 6)] to explore functions of candidate regulatory regions in a limited way. Sequences bending DNA (45-47), or phasing nucleosomes and other transcription factors (48-50) are left unaltered using BACs compared to characteristics of the gene region found endogenously. Bringing exogenous pieces of DNA together to create artificial junctions in small plasmids to trans-activate reporter genes does not adequately address the endogenous role of the regulatory sequence; and this is avoided using BACs. It is probably worth noting that DNA structure surrounding regulatory factor binding sites have evolved over long periods, and these are also left unaltered here. We noted earlier that there were 52 sites with six or more consecutive A-residues, known to cause bends in unpackaged DNA (46), in the ~28 kb upstream regulatory sequence identified as important in regulating APPb (14).

2.3 Substitution of Lox sites flanking BAC inserts with Lox66

In a different application Tn10 transposons carrying 'arm' mutants of loxP, such as lox66, have been used to successfully substitute the loxP site at one end of the genomic insert DNA in a BAC (39). It should be possible also to substitute the lox511 site at the other end of insert DNA in a similar manner using a lox site with an identical 'spacer' as lox511 but carrying a different 'arm' sequence. The usefulness of such substitutions of loxP or lox511 sites in BACs with lox66 is realized from the fact that the integration of loxP plasmids to a loxP site on the chromosome is reversible, with the equilibrium favored for excision. Targeting lox66 substituted BACs specifically to pre-positioned lox71 sites in chromosomes via lox-Cre recombination should be irreversible. Such strategies are likely to be particularly useful in systems such as the zebrafish, where classical "knock-in" technology cannot be used because of genome duplication in an ancestral teleost [see reference (39) for discussion].

2.4 Pros and cons of the transposon retrofitting approach

The insertions of exogenous DNA cassettes into BAC DNA are performed using Tn10 transposons carrying loxP or lox511 sites, and do not rely upon sequences existing in the genomic inserts of BACs. Consequently, the Tn10 transposons developed for a particular objective are applicable to all BACs in the public domain. This feature is unlike the targeting vectors used in homologous recombination dependent strategies, where the targeting vector plasmids need to be constructed anew for each BAC.

Because the insertions of these Tn10 transposons into genomic DNA in BACs appear to be largely random [see references (16) for discussion], large collections of BACs with ends progressively trimmed are generated in a single experiment. Such libraries, when made with Tn10 transposons carrying enhancer-traps or iTol2 cassettes, are uniquely suited for functionally mapping long-range gene regulatory sequences using transgenic animals. Thus multiple BACs from a contig spanning a genetic locus should enable such functional analyses to be extended over large sections of the genome. The approach does not require selecting sequences for mutational analysis to test their gene regulatory potential, thus it is an unbiased approach. It enables enhancer-trap containing BACs, or BACs retrofitted with EGFP cassettes by sequence homology based recombination, to be converted into deletion libraries with integrated iTol2 ready for chromosome integration. Thus compared to traditional approaches for enhancer-trapping used with whole genomes in animals (51-59), our approach using individual BACs, has the potential to allow a more uniform coverage of the genome because the baseline efficiency of trap insertion is reset for individual BACs in the bacterial host. Although there appear to be vast regions of the genome refractory to enhancer-trapping by traditional means, to date we have not encountered BACs refractory to Tn10 transposon insertions.

The absence of undesirable rearrangement of insert DNA due to the high content of sequence repeats in certain BACs is an additional advantage with our approach. We note that vertebrate DNA in general and mammalian DNA in particular, has large numbers of DNA sequences that are repeated (16). Use of radioactive isotopes is also avoided because no Southern blotting is required in retrofitting BACs with transposons.

A notable drawback of the transposon based approach appears to stem from the very feature that makes it readily doable: the P1 headful packaging strategy used to isolate the functionalized BAC so easily also limits the size of BAC DNA that can be analyzed to ~ 110 kb. Although the genomic DNA insert is truncated in the process, the resulting 103 kb insert DNA size, the remainder being modified BAC vector, is unlikely to be a disadvantage in most applications; because a majority of vertebrate genes can be housed in their entirety within this size limit. Almost half of evolutionarily conserved non-coding gene-regulatory sequences in vertebrate genomes (60), and probably a similar fraction of those that are conserved in function and shape but not in sequence (43, 44), are located within this span of DNA adjoining start sites of transcription of genes.

Lastly, both the lox sites flanking insert DNA in BACs can be readily substituted with lox sites with identical spacer but different arms using Tn10 transposons (39). Such replacements would make the ~110 kb retrofitted BACs amenable to "knock-in" strategies of a slightly different type: BACs with lox66 substituted for loxP should be targetable to lox71 sites in chromosomes of organisms such as zebrafish, a system where conventional homologous recombination-mediated "knock-in" technology is unavailable due to genome duplication in an ancestral teleost during evolution (61).

Clearly the transposon retrofitting strategy and those based on homologous recombination have strengths that appear somewhat complimentary in nature. Thus a judicious approach might be to use a combination of the two methodologies, as demonstrated in earlier studies (40, 15). Introducing the Tol2 inverted repeat ends to the ends of genomic DNA insert in a BAC that was previously functionalized with a reporter gene appears straightforward and easier to do with the Tn10 transposon approach. The libraries of iTol2 inserted BACs generated should also facilitate integration of trimmed single genes into the germline and help to functionally map *cis*-acting gene regulatory sequences in animals. The approach should be applicable to a wider variety of BACs, including those with sequence repeats.

3. List of abbreviations

BAC- bacterial artificial chromosome/ **FIGE**- Field inversion gel electrophoresis/ **APPb**-amyloid precursor protein gene b/ **EGFP**-enhanced green fluorescent protein

4. Acknowledgements

The project described was supported by Award Number P20MD000175 from the National Center on Minority Health and Health Disparities (NCMHD) and funds from the North Carolina Biotechnology Center. The content is solely the responsibility of the authors and does not necessarily represent the official views of the NCMHD or the National Institutes of Health. We thank Shanta Mackinnon and Charles Hatcher for zebrafish eggs, and Ms. Rosalind Grays, Connie Keys, Crystal McMichael and Darlene Laws for support and encouragement. PKC would like to thank Drs. Ken Harewood and Sean Kimbro for encouragement and support.

5. References

[1] Yang XW, Model P, Heintz N. (1997). Homologous recombination based modification in Escherichia coli and germline transmission in transgenic mice of a bacterial artificial chromosome. *Nat Biotechnol* 9: 859-865

[2] Zhang Y, Buchholz F, Muyrers JP, Stewart AF. (1998). A new logic for DNA engineering using recombination in Escherichia coli. *Nat Genet.* 20: 123-128.

[3] Jessen JR, Meng A, McFarlane RJ, Paw BH, Zon LI, Smith GR, Lin S. (1998). Modification of bacterial artificial chromosomes through chi-stimulated homologous recombination and its application in zebrafish transgenesis. *Proc Natl Acad Sci U S A*. 95: 5121-5126.

[4] Muyrers JP, Zhang Y, Testa G, and Stewart AF (1999). Rapid modification of Bacterial Artificial Chromosomes by ET recombination. *Nucleic Acids Res 27:* 1555-1557.

[5] Gong S, Yang XW, Li C, Heintz N. (2002) Highly efficient modification of bacterial artificial chromosomes (BACs) using novel shuttle vectors containing the R6Kgamma origin of replication. *Genome Res.* 12: 1992-1998.

[6] Warming S, Costantino N, Court DL, Jenkins NA, Copeland NG (2005) Simple and highly efficient BAC recombineering using galK selection. *Nucleic Acids Res. 33: e36.*

[7] Yang Z, Jiang H, Chachainasakul T, Gong S, Yang XW, Heintz N, Lin S. (2006). Modified bacterial artificial chromosomes for zebrafish transgenesis. *Methods.* 39:183-188.

[8] Mortlock DP, Guenther C, Kingsley DM. (2003) A general approach for identifying distant regulatory elements applied to the Gdf6 gene. *Genome Res.* 9:2069-2081.

[9] Jessen, J. R., Willett, C. E., and Lin, S. (1999) Artificial chromosome transgenesis reveals long-distance negative regulation of rag1 in zebrafish. *Nat Genet.* 23: 15-16.

[10] Carvajal, J. J., Cox, D., Summerbell, D., and Rigby, P. W. (2001) A BAC transgenic analysis of the Mrf4/Myf5 locus reveals interdigitated elements that control activation and maintenance of gene expression during muscle development. *Development.* 128: 1857-1868.

[11] Orford M, Nefedov M, Vadolas J, Zaibak F, Williamson R, Ioannou PA. (2000) Engineering EGFP reporter constructs into a 200 kb human beta-globin BAC clone using GET Recombination. *Nucleic Acids Res* 18: E84

[12] Chatterjee, P.K., Coren, J.C. (1997). Isolating large nested deletions in bacterial and P1 artificial chromosomes by *in vivo* P1 packaging of products of Cre-catalyzed recombination between the endogenous and a transposed *lox*P site. *Nuc. Acids Res.* 25: 2205-2212.

[13] Chatterjee, P.K., Yarnall, D.P., Haneline, S.A., Godlevski, M.M., Thornber , S.J., Robinson, P.S., Davies, H.E., White, N.J., Riley, J.H. and Shepherd, N.S. (1999). Direct Sequencing of Bacterial and P1 Artificial Chromosome Nested-deletions for Identifying Position-Specific Single Nucleotide Polymorphisms. *Proc. Natl. Acad. Sci. (USA)* 96: 13276-13281.

[14] Shakes, L.A., Malcolm, T.L., Allen, K.L., De, S., Harewood, K.R., & Chatterjee, P.K. (2008). Context dependent function of APPb Enhancer identified using Enhancer Trap-containing BACs as Transgenes in Zebrafish. *Nucleic Acids Research* 36: 6237-6248.

[15] Shakes LA, Abe G, Eltayeb MA, Wolf HM, Kawakami K and Chatterjee PK (2011). Generating libraries of iTol2-end insertions at BAC ends using loxP and lox511 Tn10 transposons. *BMC Genomics* 12: 351

[16] Gilmore, R.C., Baker Jr., J., Dempsey, S., Marchan, R., Corprew Jr., R.N.L., Maeda, N., Smithies, O., Byrd, G., Bukoski, R.D., Harewood, K.R. & Chatterjee, P.K. (2001). Using PAC nested-deletions to order contigs and microsatellite markers at the high repetitive sequence containing Npr3 gene locus. *Gene* 275: 65-72.

[17] Craig, NL. (1997). Target site selection in transposition. *Annu Rev Biochem* 66: 437-474.

[18] Chatterjee, PK, and Briley, LP (2000) Analysis of a Clonal Selection Event during Transposon-Mediated Nested-Deletion Formation in Rare BAC and PAC Clones. *Analytical Biochemistry*, 285: 121-126.

[19] Chatterjee, P.K. (2004). Retrofitting BACs and PACs with LoxP Transposons to Generate Nested Deletions. "Bacterial Artificial Chromosomes" vol 1, pp 231-241. *Methods in Mol. Biology* series vol 255, The Humana Press Inc. Editors: Shaying Zhao & Marvin Stodolsky.

[20] Yarmolinsky, M.B. & Sternberg N. (1988). Bacteriophage P1 in *The Bacteriophages,* Vol 1 edited by Richard Calendar, *Plenum Publishing Corporation.*

[21] Chatterjee, P. K., Shakes, L. A., Srivastava, D.K., Garland, D.M., Harewood, K.R., Moore, K.J., and Coren, J.S. (2004). Mutually Exclusive Recombination of Wild Type and Mutant *lox*P Sites *in vivo* Facilitates Transposon-Mediated Deletions from Both Ends of Genomic DNA in PACs. *Nucleic Acids Research* 32: 5668-5676.

[22] Chatterjee, P.K. and Sternberg, N.L. (1996). Retrofitting high molecular weight DNA cloned in P1: introduction of reporter genes, markers selectable in mammalian cells and generation of nested deletions. *Genet Anal Biomol. Eng.* 13: 33-42.

[23] Chatterjee, P.K., Mukherjee, S., Shakes, L.A., Wilson, III, W., Harewood, K.R. & Byrd, G. (2004). Selecting Transpositions of a Markerless Transposon Using Phage P1 Headful Packaging: New Transposons for Functionally Mapping Long Range Regulatory Sequences in BACs. *Analytical Biochemistry* 335: 305-315.

[24] Shakes, L. A., Garland, D.M., Srivastava, D.K., Harewood, K.R. and Chatterjee, P. K. (2005). Minimal Cross-recombination between wild type and loxP511 sites *in vivo*

facilitates Truncating Both Ends of Large DNA Inserts in pBACe3.6 and Related Vectors. *Nucleic Acids Research.* 33: *e118.*

[25] Chatterjee, P.K. and Baker Jr. J. C. (2004). Preparing Nested Deletions Template DNA for Field Inversion Gel Electrophoresis Analyses and Position-Specific End Sequencing With Transposon Primers. "Bacterial Artificial Chromosomes" vol 1, pp 243-254. *Methods in Mol. Biology* series vol 255, The Humana Press Inc. Editors: Shaying Zhao & Marvin Stodolsky.

[26] Frengen E, Weichenhan D, Zhao B, Osoegawa K, van Geel M, de Jong PJ (1999). A modular, positive selection bacterial artificial chromosome vector with multiple cloning sites. *Genomics* 58: 250-253.

[27] Osoegawa K., Woon P.Y., Zhao B., Frengen E., Tateno M., Catanese J.J. de Jong P.J. (1998). An improved approach for construction of bacterial artificial chromosome libraries *Genomics* 52:1-8.

[28] Osoegawa K, Tateno M, Woon PY, Frengen E, Mammoser AG, Catanese JJ, Hayashizaki Y, de Jong PJ. (2000). Bacterial artificial chromosome libraries for mouse sequencing and functional analysis. *Genome Res.* 10: 116-128.

[29] Osoegawa K, Mammoser AG, Wu C, Frengen E, Zeng C, Catanese JJ, de Jong PJ. (2001). A bacterial artificial chromosome library for sequencing the complete human genome. *Genome Res.* 11: 483-496.

[30] Osoegawa K, Zhu B, Shu CL, Ren T, Cao Q, Vessere GM, Lutz MM, Jensen-Seaman MI, Zhao S, de Jong PJ (2004). BAC resources for the rat genome project. *Genome Res.* 4:780-785.

[31] Krzywinski M, Bosdet I, Smailus D, Chiu R, Mathewson C, Wye N, Barber S, Brown-John M, Chan S, Chand S, Cloutier A, Girn N, Lee D, Masson A, Mayo M, Olson T, Pandoh P, Prabhu AL, Schoenmakers E, Tsai M, Albertson D, Lam W, Choy CO, Osoegawa K, Zhao S, de Jong PJ, Schein J, Jones S, Marra MA. (2004). A set of BAC clones spanning the human genome. *Nucleic Acids Res.* 32: 3651-3660.

[32] Hoess RH, Wierzbicki A, Abremski K (1986). The role of the *loxP* spacer region in P1 site-specific recombination. *Nucleic Acids Res.* 14: 2287–2300.

[33] Lee G. and Saito,I., (1998) Role of nucleotide sequences of *loxP* spacer region in Cre-mediated recombination. *Gene* 216: 55–65

[34] Langer SJ, Ghafoori AP, Byrd M, Leinwand L.(2002) A genetic screen identifies novel non-compatible loxP sites. *Nucleic Acids Res. 30: 3067-3077.*

[35] Missirlis PI, Smailus DE, Holt RA (2006). A high-throughput screen identifying sequence and promiscuity characteristics of the *lox*P spacer region in Cre-mediated recombination. *BMC Genomics.* 7: 73–85.

[36] Sheren J, Langer SJ, Leinwand LA (2007) A randomized library approach to identifying functional lox site domains for the Cre recombinase. *Nucleic Acids Res.* 35: 5464-5473.

[37] Siegel R.W., Jain,R. and Bradbury,A. (2001) Using an *in vivo* phagemid system to identify non-compatible *loxP* sequences. *FEBS lett.* 505: 467–473.

[38] Wang Z., Engler,P., Longacre,A. and Storb,U. (2001) An efficient method for high-fidelity BAC/PAC retrofitting with a selectable marker for mammalian cell transfection. *Genome Res.* 11: 137–142.

[39] Chatterjee, PK, Shakes, LA, Stennett, N, Richardson, VL, Malcolm, TL & Harewood, KR. (2010). Replacing the wild type *loxP* site in BACs from the public domain with *lox66* using a *lox66* transposon. *BMC Res Notes 3: 38*

[40] Chi, X., Chatterjee, P.K., Wilson III, W., Zhang, S-X., DeMayo, F., and Schwartz, R.J. (2005). Complex Cardiac Nkx2-5 Gene Expression Activated By Noggin Sensitive Enhancers Followed By Chamber Specific Modules. *Proc. Natl. Acad. Sci. (USA)* 102: 13490-13495.

[41] Brake, R.L., Chatterjee, P.K., Kees, U.R & Watt, P.M. (2004). The functional mapping of long-range transcription control elements of the *HOX11* proto-oncogene. *Biochem. Biophys. Res. Com.* 313: 327-335.

[42] Suster ML, Sumiyama K, Kawakami K (2009) Transposon-mediated BAC transgenesis in zebrafish and mice. *BMC Genomics* 10:477

[43] Fisher S, Grice EA, Vinton RM, Bessling SL, McCallion AS (2006) Conservation of RET regulatory function from human to zebrafish without sequence similarity *Science 14: 276-279.*

[44] Parker SC, Hansen L, Abaan HO, Tullius TD, Margulies EH. (2009) Local DNA topography correlates with functional noncoding regions of the human genome. Science. Apr 17; 324 (5925): 389-92. Epub 2009 Mar 12.

[45] Fried MG and Crothers DM (1983) CAP and RNA polymerase interactions with the lac promoter: binding stoichiometry and long range effects. *Nucleic Acids Res.* 11: 141–158

[46] Ulanovsky L, Bodner M, Trifonov E.N., Choder M.(1986) Curved DNA: design, synthesis, and circularization. *Proc Natl Acad Sci U S A.* 83: 862-866

[47] Gartenberg M.R., Ampe C, Steitz T.A., Crothers D.M. (1990) Molecular characterization of the GCN4-DNA complex *Proc Natl Acad Sci U S A.* 87: 6034-6038

[48] Hebbar P.B., Archer T.K. (2007) Chromatin-dependent cooperativity between site-specific transcription factors *in vivo. J Biol Chem* 282: 8284-8291.

[49] Ganapathi M, Singh G.P., Sandhu K.S., Brahmachari S.K, Brahmachari V. (2007) A whole genome analysis of 5' regulatory regions of human genes for putative cis-acting modulators of nucleosome positioning. *Gene.* 391: 242-251.

[50] S Hardy and T Shenk (1989) E2F from adenovirus-infected cells binds cooperatively to DNA containing two properly oriented and spaced recognition sites. *Mol Cell Biol* 9: 4495–4506.

[51] O'Kane CJ, Gehring WJ. (1987). Detection in situ of genomic regulatory elements in Drosophila. *Proc Natl Acad Sci U S A.* 84: 9123-9127.

[52] Wilson C, Pearson RK, Bellen HJ, O'Kane CJ, Grossniklaus U, Gehring WJ. (1989). P-element-mediated enhancer detection: an efficient method for isolating and characterizing developmentally regulated genes in Drosophila. *Genes Dev.* 3: 1301-1313.

[53] Korn R, Schoor M, Neuhaus H, Henseling U, Soininen R, Zachgo J, Gossler A. (1992). Enhancer trap integrations in mouse embryonic stem cells give rise to staining patterns in chimaeric embryos with a high frequency and detect endogenous genes. *Mech Dev.* 39: 95-109.

[54] Grabher C, Henrich T, Sasado T, Arenz A, Wittbrodt J, Furutani-Seiki M. (2003). Transposon-mediated enhancer trapping in medaka. *Gene.* 322: 57-66.

[55] Balciunas D, Davidson AE, Sivasubbu S, Hermanson SB, Welle Z, Ekker SC. (2004). Enhancer trapping in zebrafish using the Sleeping Beauty transposon. *BMC Genomics*. 5: 62.

[56] Kawakami, K., Takeda, H., Kawakami, N., Kobayashi, M., Matsuda, N., and Mishina, M. (2004) A transposon-mediated gene trap approach identifies developmentally regulated genes in zebrafish. *Developmental Cell* 7, 133-144

[57] Kawakami, K. (2005) Transposon tools and methods in zebrafish. *Developmental Dynamics* 234, 244-254

[58] Ellingsen S, Laplante MA, Konig M, Kikuta H, Furmanek T, Hoivik EA, Becker TS (2005) Large-scale enhancer detection in the zebrafish genome *Development*. 132: 3799-3811.

[59] Nagayoshi, S., Hayashi, E., Abe, G., Osato,N., Asakawa, K., Urasaki, A., Horikawa, K., Ikeo, K., Takeda, H., and Kawakami, K. (2008) Insertional mutagenesis by the Tol2 transposon-mediated enhancer trap approach generated mutations in two developmental genes: tcf7 and synembryn-like. *Development* 135(1), 159-169

[60] Woolfe A, Goodson M, Goode DK, Snell P, McEwen GK, Vavouri T, Smith SF, North P, Callaway H, Kelly K, Walter K, Abnizova I, Gilks W, Edwards YJ, Cooke JE, Elgar G. (2005) Highly conserved non-coding sequences are associated with vertebrate development. *PLoS Biol.* (2005) Jan; 3: e7.

[61] Kleinjan DA, Bancewicz RM, Gautier P, Dahm R, Schonthaler HB, Damante G, Seawright (2008). A, Hever AM, Yeyati PL, van Heyningen V, Coutinho P. Subfunctionalization of duplicated zebrafish pax6 genes by cis-regulatory divergence. *PLoS Genet.* 4: e29

5

Functional Profiling of Varicella-Zoster Virus Genome by Use of a Luciferase Bacterial Artificial Chromosome System

Lucy Zhu and Hua Zhu
UMDNJ-New Jersey Medical School
United States

1. Introduction

The varicella-zoster virus (VZV or Human Herpesvirus-3) is a member of the human herpesvirus family. Identified as the causative agent of chickenpox and shingles, VZV is a significant pathogen in the United States, infecting over 90% of the US population (Abendroth & Arvin, 1999). Primary VZV infections generally occur during childhood and result in a relatively benign illness termed varicella (chickenpox). However, like all herpesviruses, VZV will establish latency in the host's sensory neurons. This occurs specifically in the trigeminal ganglia and dorsal root ganglia (Arvin, 1996; Gilden et al., 2000), where the virus can hide from the immune system for years and often a lifetime.

Reactivation of this virus with increasing age or immunosuppression results in herpes zoster (shingles). Herpes zoster is a more painful, neurocutaneous infection associated with acute or chronic neuropathic pain and a significant incidence of post-herpetic neuralgia, a complication causing many patients to continue suffering excruciating pain, lasting anywhere from months to years after infection due to the resulting nerve damage (Cohen et al, 2007). This not only greatly increases the cost of medical care, but also significantly compromises the quality of life in the elderly (Pickering & Leplege, 2010; Opstelten et al., 2010; Burgoon et al, 1957).

1.1 The varicella vaccine

In the early 1970s, Japanese researchers isolated a VZV sample from the blood of a small boy. Through serial passage in cell culture, scientists were able to successfully develop the first live attenuated varicella vaccine (Takahashi et al., 1974; Arvin, 2001; Gershon, 2001). The vaccine strain was termed the Japanese Oka varicella virus (v-Oka). By 1995, this chickenpox vaccine was introduced to the United States and quickly recommended for vaccination in all children. Since then, chickenpox incidence in the US has dramatically declined; the effectiveness of this vaccine is estimated to be between 72% and 98% (Hambleton & Gershon, 2005). Nevertheless, outbreaks of chickenpox are still ever-present. Furthermore, the vaccine may indirectly increase the occurrence of herpes zoster in the elderly population by lessening the number of natural infections, and therefore lowering the exposure to wild-type VZV that would boost natural immunity (Arvin, 2001; Galea et al,

2008; Volpi, 2005). It is also important to note that the current shingles vaccine has been shown to only reduce the risk of shingles by 50% (Oxman et al., 2005). Because of this, VZV continues to be an important public health concern. In order to improve future prevention and treatment of VZV infections, a better understanding of VZV's biology and pathogenesis is critical.

1.2 VZV research methods

VZV contains the smallest genome among the eight human herpesviruses, consisting of a 125-kb double-stranded DNA genome that encodes 70 unique open reading frames (ORFs). The function of most of these ORFs, however, was largely unknown until recent years. This is in part due to the absence of both a genetic tool to efficiently generate mutant clones for loss-of-function studies and a true animal model for large scale screening of *in vivo* virulence factors (Cohen et al., 2007).

Obstacles in mutagenizing VZV include its large genome size, narrow host range, and marked differences in replication cycles when studied *in vitro* versus *in vivo* (Arvin, 1996; Cohen, 2001). A once prevalent technique to create recombinant VZV variants was the four-cosmid system, made by cloning overlapping segments of the VZV genome into four large cosmids (Cohen & Seidel, 1993; Mallory et al, 1997; Niizuma et al., 2003). Co-transfection of these cosmids, one of which containing a mutation in the desired ORF, created a recombinant VZV variant. However, this method alone faced many challenges. For example, research was thwarted because co-transfection of the large cosmids into permissive mammalian cells and multiple homologous recombination events within a single cell were necessary to generate the full-length viral genome (Zhang et al, 2008).

More recent developments have helped to circumvent these problems by cloning the entire VZV genome as a bacterial artificial chromosome (VZV_{BAC}) (Nagaike et al., 2004). This approach provides easy and efficient manipulation of the viral genome and rapid isolation of recombinant viruses, making the systemic deletion of every ORF in the genome feasible. A firefly luciferase cassette is also inserted into the VZV_{BAC} to produce a novel luciferase VZV_{Luc} BAC. This allows us to not only generate VZV ORF deletion mutants, but also monitor its subsequent growth in cultured cells.

2. Generation of a VZV_{BAC}

Viral BACs are created when a BAC vector sequence is inserted into the viral genome. Typically, a BAC vector is roughly 10-kb long and contains an origin of replication, genes essential for BAC replication, and genes to control the rate of replication; ideally the copy number should be limited to one or two BACs per bacterial cell (Warden et al., 2010). An antibiotic resistance marker and selection marker, such as a green fluorescent protein, are also added to select for bacterial colonies harboring the BAC vector and isolate these BAC-containing recombinant viruses. BAC vectors in addition must also be flanked by a 500-1000-bp sequence homologous to the target sequence at the site of insertion. Lastly, loxP sites are commonly included at both ends of the BAC sequence to excise the vector after recombinant viruses are generated, as is required for vaccine production (Zhang et al., 2007, see below).

2.1 The pUSF-6 vector

For our purposes, the VZV_{BAC} was constructed from a pUSF-6 vector. As shown in Fig. 1, this vector contains the prokaryotic replication origin (*ori*), replication and partition function (*repE, parA, parB*) genes, chloramphenicol resistance (cam^r) gene, and a green fluorescent protein (*GFP*) gene. Insertion of a GFP reporter gene in BAC DNA is a popular means to visualize *in vitro* infections in cell culture. Viral GFP is expressed using the SV40 early promoter and polyadenylation signals, which activate the gene during the appropriate stages of viral replication and cause the cell to fluoresce (Marchini et al., 2001). The vector is also flanked by two 500-bp VZV fragments and contains a loxP site at each end (Fig. 1).

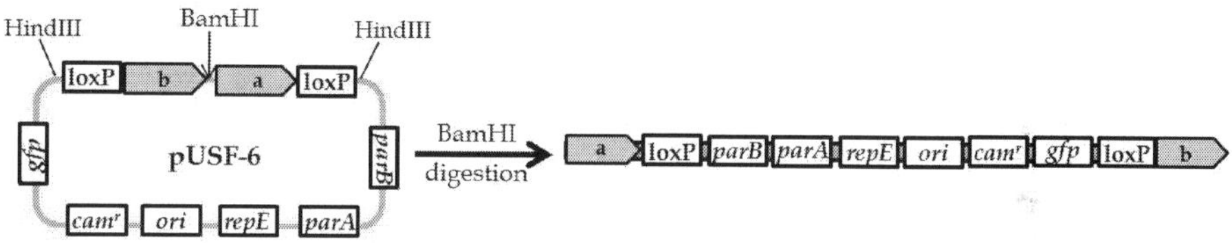

Fig. 1. The BAC vector, pUSF-6. The vector contains the prokaryotic replication origin *(ori)*, replication and partition (*repE, par,* and *parB*) genes, cam^r gene, green fluorescent protein (*GFP*) gene, two *loxP* sites and two VZV homologous sequences, a and b. To insert this BAC vector into a VZV cosmid, pUSF-6 was digested by BamHI, resulting in a linear fragment.

2.2 Construction of the VZV BAC (VZV_{BAC})

Generally, BAC vectors can be directly inserted into viral genomes via homologous recombination. However, this method cannot be used to construct the VZV_{BAC} because the virus' highly cell-associated nature makes isolation of the VZV genome and purification of recombinant plaques difficult (Nagaike et al., 2004). Instead, VZV_{BAC} clones are constructed using a set of four overlapping cosmids spanning the entire VZV genome (Fig. 2).

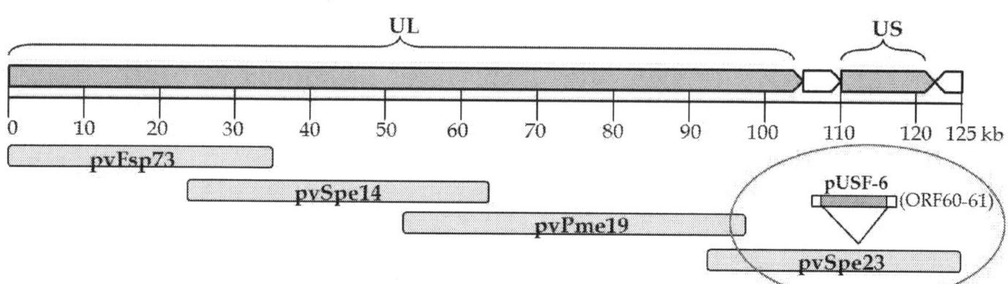

Fig. 2. Schematic diagram the VZV genome and four-cosmid system. The VZV clinical strain, p-Oka, consists of a 125-kb genome with unique long (UL) and unique short (US) segments. Four cosmids with overlapping VZV genomic segments are shown. The BAC vector-containing plasmid, pUSF-6, was inserted between ORF60-61.

First, the pUSF-6 vector was inserted into VZV cosmid pvSpe23, between ORF60 and ORF61 (Fig. 2), via homologous recombination (Yu et al., 2000). The BAC-containing cosmid was then co-transfected with three complementary cosmids (Niizuma et al., 2003) into human melanoma (MeWo) cells. Because each of the four cosmids contains an overlapping sequence, the cosmids can recombine into one large circular virus genome via homologous recombination to create the recombinant virus. Our use of the GFP selectable marker in the pUSF-6 vector allowed for visualization of the recombinant VZV_{BAC} in plaques that form post-transfection (Fig. 3C). Finally, the VZV_{BAC} DNA from the infected cells was purified and transformed into *E. coli*. Chloramphenicol-resistant colonies were selected for and used to isolate the desired VZV_{BAC} DNA (Zhang et al., 2007).

We used restriction enzyme digestion and DNA sequencing to verify the integrity and stability of the VZV_{BAC} DNA. VZV_{BAC} DNA was digested by HindIII, yielding the predicted digestion pattern with a sum of ~130 kb (Fig. 3G), thus indicating that no large deletions and rearrangements were present. In addition, the ORF62/71 gene was sequenced to check for base-pair changes in the VZV_{BAC} genome after synthesis in *E. coli*. This large duplicated gene, encoding an immediate-early transactivating protein (Perera et al, 1992), was amplified via PCR and cloned into a pGEM-T vector for sequencing. Because the ORF62/71 sequences in the VZV_{BAC} were identical to those of the published p-Oka strain, we can conclude that the BAC DNA in *E. coli* is stable.

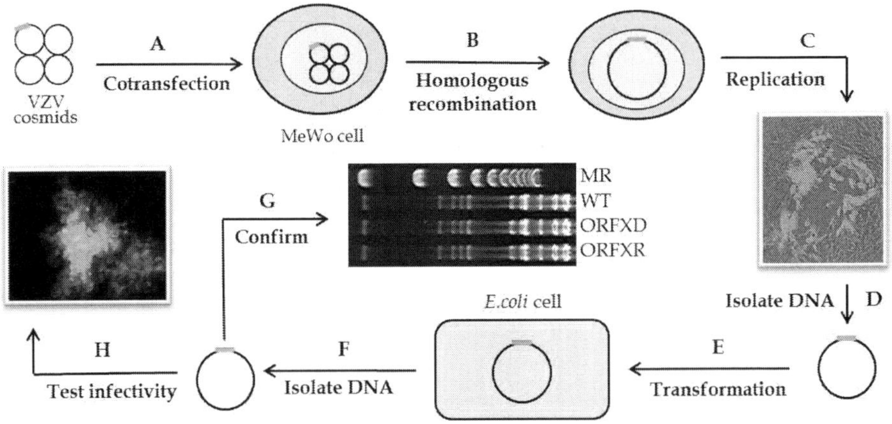

Fig. 3. Construction of VZV BAC. (A) The BAC-containing cosmid was co-transfected with the three complementary cosmids into MeWo cells. (B) Homologous recombination between cosmids formed a circular, full-length VZV_{BAC} genome. (C) The recombinant BAC was replicated, and produced a plaque visualized with the GFP marker. (D) Circular DNA was isolated from infected cells, (E) transformed into *E. coli*, and selected for cm^R colonies. (F) The VZV_{BAC} DNA was isolated from *E. coli* and (G) verified by restriction digestion and partial sequencing. (H) The infectivity and integrity of the VZV_{BAC} were tested by transfecting BAC DNA into MeWo cells to generate the VZV virus.

The GFP marker in the viral genome was also tested; MeWo cells were infected with VZV_{BAC} and continuously passed four times (1:10 dilution) over two weeks. On examination of the plaques under a fluorescent microscope, all VZV_{BAC}-infected cells fluoresced green, signifying the stability of the GFP marker in the viral genome. Lastly, the infectivity and integrity of the VZV_{BAC} were confirmed by transfecting BAC DNA into MeWo cells to produce the virus. A summary of the process to construct a VZV BAC and verify its integrity is illustrated in Fig. 3.

3. Application of VZV_{BAC} with a luciferase marker

Visualization markers are often inserted into the viral BAC genome to detect and quantify viral replication. Two methods frequently utilized are fluorescence imaging and bioluminescence imaging. Fluorescence-based imaging, such as through the use of a GFP reporter gene, is a common method to monitor *in vitro* infections and allows researchers to study the interaction of a given virus with its host (Tang, 2008). Visualization of *in vivo* infections on the other hand, can be established by the use of bioluminescence imaging (BLI).

3.1 Background on bioluminescence imaging

Developed over the last decade, bioluminescence imaging is a technology that enables visualization of viral gene expression in live tissues and animals (Tang et al, 2008). With little surprise, BLI has become a powerful technique for studying VZV pathogenesis.

Bioluminescence is the production of light by living organisms, resulting from a chemical reaction in which chemical energy is converted to light energy (Hastings, 1983; Kurfurst et al., 1983). BLI systems generate bioluminescence using two compounds - luciferase and its substrate luciferin. Luciferase is a class of enzymes commonly exploited as a reporter gene for transcriptional regulation studies (Doyle et al., 2004). Most extensively employed is the luciferase of the North American firefly (*Photinus pyralis*), which can be expressed in mammalian cells by inserting the gene under the control of a promoter. Because firefly luciferase generates a bioluminescence wavelength that can efficiently penetrate tissues, it serves as an excellent indicator for *in vivo* studies (Tang et al, 2008).

There are many advantages to employing BLI over other bio-imaging techniques; one of the key factors is its use of luciferin. Not only can luciferin permeate all tissues *in vivo* (including cell membranes and the blood-brain barrier) (Contag et al., 1997; Rehemtulla et al., 2002), the substrate can also be administered numerous times to the same animal and provides great accuracy, due to its low toxicity and high sensitivity, respectively (Contag et al., 1997; Rehemtulla et al., 2000). When exposed to the appropriate luciferin substrate, luciferase will catalyze an oxidation reaction to produce light visible to the human eye. The light's intensity depends on the amount of luciferase present, which can be determined by quantifying the relative amounts bioluminescence emitted *in vivo* via computer-based analysis. For this reason, engineering viral BACs to express luciferase can be especially valuable for monitoring the activities of the promoter that mediates gene regulation, detecting sites of viral infections, and quantifying viral replication in living cultures and animals (Zhang et al., 2007; Zhang et al., 2008; Dulal et al., 2009; Zhang et al., 2010).

3.2 The VZV_{BAC} with a luciferase marker (VZV_{Luc})

To generate a VZV strain expressing luciferase, a firefly luciferase expression cassette was inserted into the intergenic region between ORF65 and ORF66 of the VZV_{BAC} genome. This clone was transfected into MeWo cells to produce the VZV_{Luc} strain. 24 hours later, cell culture media was replaced with media containing 150 μg/ml D-luciferin. After incubation at 37°C for 10 minutes, bioluminescent signals were observed and quantified using an *In Vivo* Imaging System (IVIS).

Upon analysis (Fig. 4), the growth of VZV_{Luc} closely resembled that of its parental VZV_{BAC} (data not shown). This confirms that the addition of a luciferase reporter to the viral BAC did not change its growth properties. Unlike its parental strain however, only cells infected with VZV_{Luc} expressed high levels of luciferase activity (Fig. 4A) and emitted a strong bioluminescence after the addition of D-luciferin (Fig. 4B).

To explore the possibility of using bioluminescent signals as an indicator of viral growth, bioluminescent assays were compared to the conventional infectious center assay. Plates were inoculated with wild-type (WT) VZV and VZV_{Luc}. Their viral titers were quantified daily via both methods for seven days and the data collected was used to construct viral growth curves (Fig. 4C). As the figure illustrates, the intensity of the bioluminescent signals strongly correlated with the viral titers generated by an infectious center assay. Thus, this data supports BLI as an alternative method for growth curve assays and quantifying viral titers.

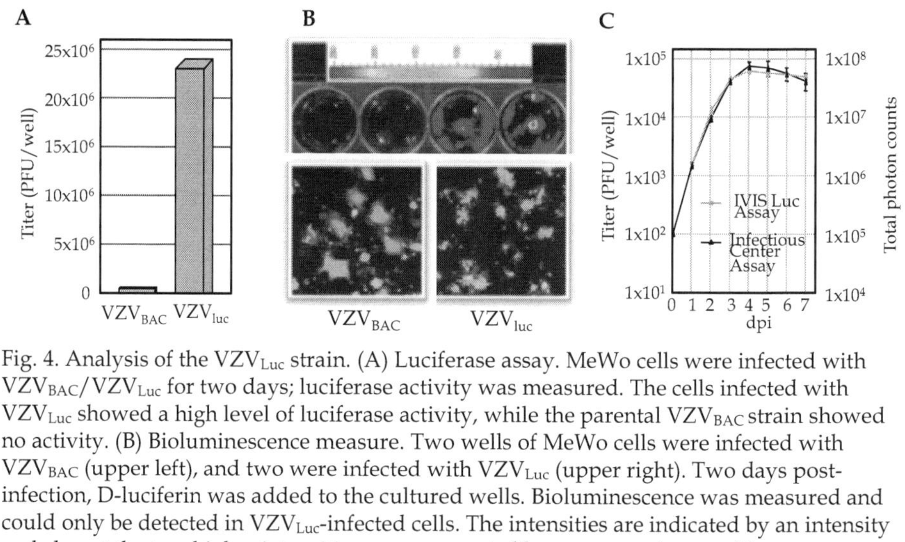

Fig. 4. Analysis of the VZV_{Luc} strain. (A) Luciferase assay. MeWo cells were infected with VZV_{BAC}/VZV_{Luc} for two days; luciferase activity was measured. The cells infected with VZV_{Luc} showed a high level of luciferase activity, while the parental VZV_{BAC} strain showed no activity. (B) Bioluminescence measure. Two wells of MeWo cells were infected with VZV_{BAC} (upper left), and two were infected with VZV_{Luc} (upper right). Two days post-infection, D-luciferin was added to the cultured wells. Bioluminescence was measured and could only be detected in VZV_{Luc}-infected cells. The intensities are indicated by an intensity scale bar at the top; higher intensities are represented by warmer colors, and lower intensities are represented by cooler colors. The infection was verified by the GFP-positive plaques (bottom panel). (C) Correlation of luminescence and plaque numbers. Growth curves generated by an infectious center assay (black curve and left scale) and a bioluminescence assay (green curve and right scale) were compared.

3.3 Bioluminescence imaging for studying VZV in SCID-hu mice

Another useful application of bioluminescence imaging is the live-image analysis of VZV replication in severe combined immunodeficient mice with human tissue xenografts (SCID-hu mice). Because VZV only infects human cells, *in vivo* studies of VZV pathogenesis have been limited to the use of immunodeficient mice with human tissue implants. However, although SCID-hu mice are established as appropriate models for studying VZV pathogenesis (Besser et al, 2003; Ku et al., 2005; Zerboni et al., 2005), collecting quantitative data has been a major challenge. Since measuring viral growth required the mice to first be euthanized, it was impossible to monitor the progression of the viral infection in the same mouse. In addition, viral titers tend to vary from animal to animal because of the differently sized implants (Moffat & Arvin, 1999), thus hindering not only the frequency of data collection, but also the accuracy as well. These factors greatly impeded efforts to study large numbers of VZV variants and made it difficult to discern minor phenotypic differences leading to pathogenesis.

The development of BLI has been extremely helpful to circumvent these obstacles. Luciferase provides a visible marker for detecting VZV in human tissues within living animals. By using VZV_{Luc}, the SCID thymus-liver mouse model, and *In Vivo* Imaging System (IVIS, Xenogen), the spread of the VZV infection can be frequently monitored in the same mouse over an extended period of time; thereby, allowing the generation of credible growth curves to gain accurate insights into VZV's growth kinetics *in vivo*.

We applied this method to explore VZV replication and measure its spread *in vivo*. Human fetal thymus and liver tissue were implanted under the left kidney capsule of the SCID mouse. Over the course of the next few months, the implanted tissue developed into a thymus-like organ consisting mainly of T cells. VZV-infected cells were then inoculated into the SCID-hu mice with thymus-liver implants. VZV replication was measured *in vivo* after the injection of the luciferin substrate, using an IVIS. Each mouse was imaged daily, starting four hours after inoculation (i.e day zero), for eight days (Fig 5A).

Our data depicts the daily increase in bioluminescence emitted from the infected implants (Fig. 5B). The quantified signals were plotted to generate an *in vivo* growth curve (Fig. 5C). As shown, VZV grew rapidly in human T cells, doubling approximately every 12 hours and peaking at seven days postinfection. The exponential growth curve is then followed by a steady state where the viral infection reaches the saturation limit of the implant.

We also tested the VZV_{Luc} viruses for their spread and detection in human fetal skin xenografts *in vivo*. Similar to the process outlined above, human skin tissues were introduced into SCID mice. Four weeks after implantation, VZV_{Luc} virus was inoculated into the skin tissues and viral growth was monitored every two to three days for 15 days using an IVIS. High luciferase activity was detected in the implants (data not shown), verifying VZV_{Luc}'s ability to grow in skin tissue *in vivo*.

In short, by engineering VZV to express luciferase enzymes, bioluminescence imaging can be used to monitor the progression of viral growth and quantify viral replication in organ cultures and SCID-hu mice. Compared to the traditional infectious center assay, BLI not only saves time and labor, but also significantly increases the reproducibility of results (Doyle et al., 2004). Moreover, the presence of luciferase activity indicates viral replication in cells and not free-viral particles (Zhang et al., 2010), making BLI the most suitable method for studying this particular cell-associated virus. Consequently, the development of BLI has greatly facilitated our ability to investigate aspects of VZV infection in the SCID-hu mouse model and has significantly advanced our understanding of VZV pathogenesis and virus-cell interactions (Zerboni et al., 2010; Arvin et al, 2010; Zhang et al; 2010; Moffat & Arvin; 1999; Arvin, 2006).

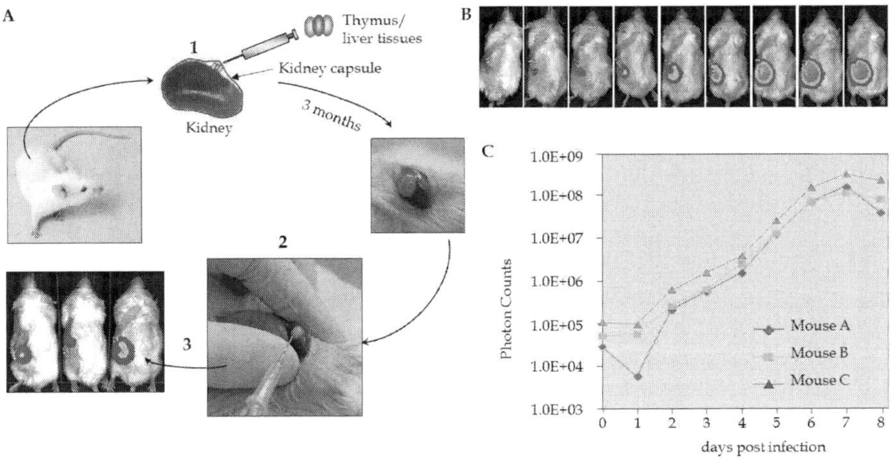

Fig. 5. Monitoring VZV_{Luc} virus replication in SCID-hu mice. (A) SCID-hu model. 1. Human fetal thymus/liver tissues were implanted under SCID mouse kidney capsule. 2. Two to three months later, the implant was inoculated with VZV_{Luc}. 3. Viral replication in human T cells was detected by IVIS. (B) Replication and progression of VZV_{Luc} in human thymus/liver implants in SCID mice. Three SCID-hu mice with thymus/liver implants were inoculated with VZV_{Luc}. Using IVIS, each mouse was scanned daily (from day 0 to day 8). Measurements were taken 10 minutes after i.p. injection with luciferin substrate. Only images from one mouse are shown. Warmer colors indicate higher viral load; colder colors indicate lower viral load. (C) VZV growth curves *in vivo*. Bioluminescence from three SCID-hu mice in the above experiment (B) was measured and VZV growth curves in human thymus/liver implants were generated.

4. Generation of recombinant VZV using a highly efficient homologous recombinant system

To test the novel VZV_{Luc} system for studying VZV pathogenesis, five single ORF deletion mutants were first generated, starting from ORF0 to ORF5, via the homologous recombination system harbored in DY380 *E. coli*. Afterwards, the VZV_{Luc} was used for genome-wide mutagenesis to systematically delete each individual VZV ORF for functional characterization of the VZV genome.

4.1 The DY380 *E. coli* strain

The DY380 *E. coli* strain offers a highly efficient homologous recombination system for chromosome engineering by enabling efficient recombination of homologous sequences as short as 40-bp (Yu et al., 2000). A defective lambda prophage supplies the function that protects and recombines linear DNA. In addition, the system is strictly regulated by a temperature-sensitive lambda repressor. This allows homologous recombination between two sequences to be transiently induced by activating the prophage through incubation at 42°C for 15 minutes.

4.2 Generation of a VZV deletion clone

The entire process to engineer a VZV ORF deletion mutant (ORFXD) is illustrated in Fig. 6. VZV_{Luc} BAC DNA was first introduced into DY380 by electroporation. Homologous recombination functions were transiently induced by increasing the culturing temperature to 42°C for 15 minutes during electroporation-competent cell preparation. A kan^R expression cassette was amplified from pGEM-oriV/kan^R by PCR using two primers containing 40-bp homologous sequences flanking the target ORF (ORFX). The PCR product was then transformed into the DY380 harboring the VZV_{Luc} BAC via electroporation. As expected, homologous recombination occured between the ORF flanking sequences of the cassette and targeted ORF, replacing the ORFX with the kan^R gene and generating an ORFXD VZV clone (Zhang et al., 2008).

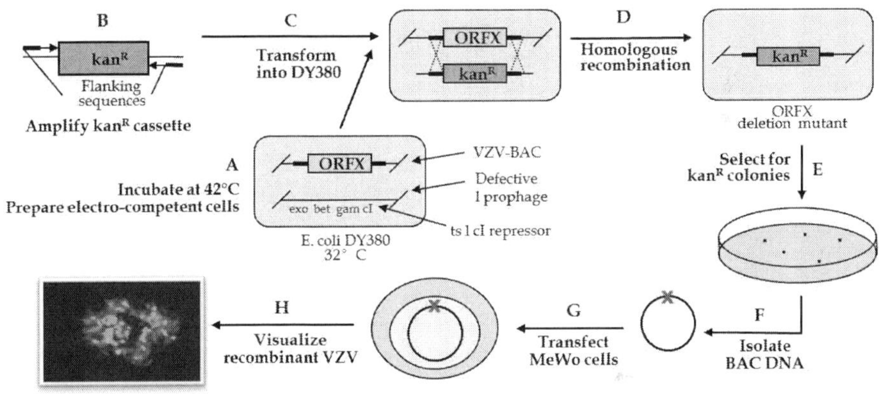

Fig. 6. Generation of a VZV deletion clone. (A) The DY380 strain permits transient induction of recombination system by incubation at 42°C for 15 min during electro-competent cell preparation. VZV_{Luc} BAC DNA was introduced into DY380 by electroporation. (B) Amplification of the kan^R expression cassette by PCR using a primer pair to add 40-bp homologous sequences flanking ORFX. (C) 200ng of the above PCR product was transformed into DY380 carrying the VZV_{Luc} BAC by electroporation. (D) Homologous recombination between upstream and downstream homologies of ORFX replaced ORFX with the kan^R cassette, creating the ORFX deletion VZV clone. (E) Recombinants were selected on LB agar plates. (F) The deletion of ORFX DNA was isolated and confirmed by testing antibiotic sensitivity and PCR analysis. The integrity of the viral genome after homologous recombination was examined by restriction enzyme digestion. (G) Purified BAC DNA was transfected into MeWo cells. (H) 3-5 days after transfection, the infected cells were visualized by fluorescence microscopy.

Successful ORF deletion clones were confirmed by three sequential procedures: 1. antibiotic sensitivity selection, 2. mini-preparation of BAC DNA with PCR verification, and 3. maxi-preparation of BAC DNA with HindIII digestion profiling. Firstly, recombinants were selected on LB plates with chloramphenicol or kanamycin for resistant colonies. It was also important to verify that the deletion clones were sensitive to ampicillin since ampicillin-resistant circular pGEM-oriV/kan^R was used as the PCR template. Multiple colonies were then selected for mini-preparation of BAC DNA to confirm the ORF deletion and kan^R replacement by PCR. Lastly the PCR-verified clones were chosen for maxi-preparation of BAC DNA and digested with HindIII to ensure that only the targeted sequence was deleted. When the digestion pattern of the deletion clone was compared to that of the parental WT VZV_{Luc} clone, no additional deletions from the genome were detectable (as shown in Fig. 3G).

Finally, to generate VZV deletion mutant viruses, these verified clones were transfected into MeWo cells, along with WT VZV_{Luc} DNA. The size and growth kinetics of the virus as measured by resultant plaques, or absence of plaques, are indicative of the essentiality of a particular VZV ORF for viral replication, discussed later.

4.3 Generation of a VZV rescue clone

VZV ORF deletion rescue clones were also generated (Fig. 7) in order to show that growth defects observed in analyses of the deletion mutants are a direct result of the deleted genes, as opposed to potential mutations in other regions of the genome. Ideally, the wild-type phenotypes should be fully restored in these rescue viruses.

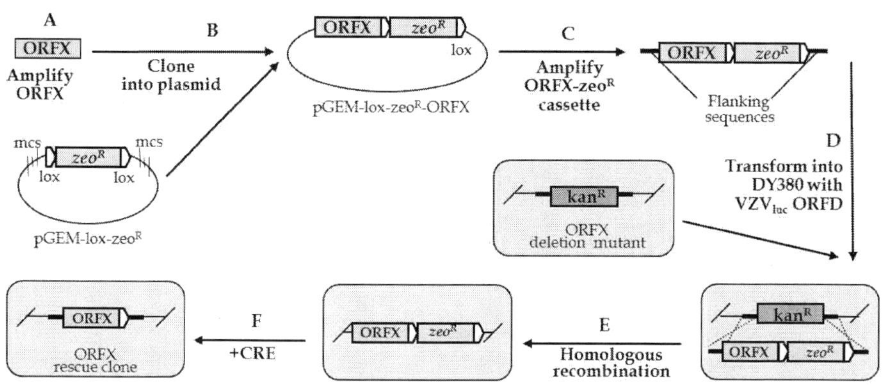

Fig. 7. Generation of a VZV rescue clone. (A) ORFX was amplified by PCR from the WT VZV BAC DNA and (B) directionally cloned into plasmid pGEM-lox-zeo to form pGEM-lox-zeo-ORFX. (C) Amplification of the ORFX-zeo^R rescue cassette by PCR using a primer pair adding 40-bp homologies flanking ORFX. (D) The PCR product was transformed into DY380 carrying the VZV_{Luc} ORFX deletion via electroporation. (E) Homologous recombination between upstream and downstream homologies of ORFX replaced kan^R with the ORFX-zeo^R rescue cassette. (F) Zeo^R and BAC vector sequences were removed post-verification by co-transfecting a Cre recombinase-expressing plasmid, creating the ORFX rescue clone.

To generate ORF deletion rescue clones (ORFXR), the targeted ORF deletion was amplified from wild-type VZV_{Luc} BAC DNA by PCR. Next, the ORFX was directionally cloned into plasmid pGEM-lox-zeo to produce pGEM-zeo-ORFX. This was then used as the template to generate the ORFX-zeo^R cassette via PCR using a primer pair to add 40-bp sequences, homologous to the kan^R cassette flanking ORFX. In a process similar to the homologous recombination system described earlier, the PCR product was transformed into DY380 carrying the ORFX deletion genome (Fig. 7C). The kan^R cassette was replaced with the ORFX-zeo^R rescue cassette by homologous recombination, thus allowing for positive selection of zeocin^R colonies. Because the zeo^R gene within the rescue cassette is flanked by two loxP sites, it can be removed along with the BAC vector, through Cre-mediated recombination, generating the desired ORFX rescue virus. Full recovery from the marked growth defects is expected after using this rescue method to restore the wild-type ORF.

5. Functional profiling of VZV genome

Even though VZV has the smallest genome among all human herpesviruses, less than 20% of the VZV genome had been functionally characterized (Cohen et al, 2007). In order to investigate VZV ORF function, we created an entire library of single VZV ORF deletion mutants using the DY380 *E. coli* recombination system. The individual functions of each ORF were determined by transfecting MeWo cells with mutant DNA, and observing the subsequent growth of viral plaques. If a VZV ORF is nonessential for viral replication, plaques corresponding to the deleted ORF should be detectable 3-5 days after transfection and resemble the growth of the wild-type virus. Plaques that grow significantly smaller and later suggest that the ORF strongly influences optimal growth. The absence of a plaque entirely implies that the ORF is essential for viral replication.

5.1 Essential VZV ORFs

The results indicate that among VZV's 70 unique ORFs, 44 ORFs are essential for viral replication in cultured MeWo cells, while 26 ORFs are nonessential (Zhang et al., 2010). Fig. 8 provides a visual representation of the entire VZV genome and categorizes the essentiality of each ORF based on the growth properties of its corresponding deletion mutant virus.

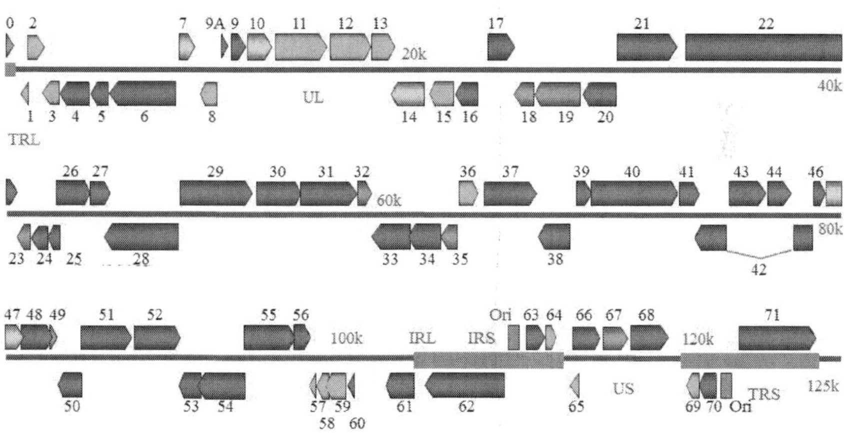

Fig. 8. VZV genome-wide functional profiling based on analysis of single viral ORF deletion mutants. Each VZV ORF is color-coded according to the growth properties of its corresponding virus gene-deletion mutant in cultured MeWo cells and human fetal skin organ cultures. The grey lines for ORF42 represent a splicing junction. For all growth curves, wild-type infections served as positive controls and mock infections served as negative controls.

Various studies, cumulatively, have found that the essential VZV ORFs encode genes for viral structural proteins, transcriptional regulatory proteins, and enzymes involved in DNA replication. The majority of these crucial ORFs encode proteins with imperative functions in maintaining the viral life cycle. For example, some ORFs are a part of the viral tegument and encode immediate-early proteins with transcriptional regulatory activity (Perera et al., 1992; Defechereux et al., 1993; Moriuchi et al., 1994). Other ORFs encode phosphoproteins primarily contained in the nuclei of infected cells (Moriuchi et al., 1993). It has also been reported that most of the VZV ORFs encoding glycoproteins also belong in this group of genes indispensable for viral replication (Mallory et al., 1998; Yamagishi et al., 2008).

Upon further analysis, we found that essential VZV genes have significantly different enrichment for functional categories than nonessential genes. As depicted by the distribution of functional annotations (Fig. 9A), essential VZV genes are significantly enriched for DNA replication and DNA packaging. These include genes encoding the subunits of VZV DNA polymerases, DNA binding proteins, DNA packaging proteins, and nucleocapsid proteins.

5.2 Nonessential VZV ORFs

As previously mentioned, 26 of the 70 unique VZV ORFs were deemed nonessential for VZV replication in MeWo cells (Zhang et al., 2010). Of these, 8 ORFs appeared to significantly affect viral growth. In viral growth assays, the peak signals from their corresponding plaques were at least 5-fold less than the peak signals from the WT parental strain. Furthermore, atypical morphology of virally infected cells, such as reduced plaque sizes and altered syncytia formation, were also frequently observed. Studies have shown that some of these ORFs affecting optimal growth encode the small and large subunit of ribonucleotide reductase (Heineman & Cohen, 1994) and specific phosphoproteins that are post-translationally modified by protein kinases (Reddy et al., 1998).

The plaques corresponding to the remaining 18 nonessential VZV ORF deletions exhibited wild-type growth in cultured MeWo cells (Zhang et al., 2010). *In vitro* growth curve analysis for viral replication showed that these ORF deletion mutants have the same growth kinetics as their wild-type parental strain, VZV_{Luc}. Nonessential genes, in general, are significantly enriched for other and unknown functional categories (Fig. 9B).

5.3 Other findings

Despite major differences between herpesvirus genomes, all the herpesviruses studied thus far have a similar number of essential genes, but varying number of nonessential genes. For example, our study suggests that the VZV genome encodes 44 essential genes and 26 nonessential genes. The herpes simplex virus 1 (HSV-1) genome encodes 37 essential genes and 48 nonessential genes (Roizman et al., 2007). Similarly, the human cytomegalovirus (HCMV), one of the largest human DNA viruses, has a genome that encodes 45 essential genes and 118 nonessential genes (Dunn et al, 2003). Furthermore, 26 of the 44 essential VZV genes have essential homologues in HSV and 18 of the 44 have essential gene homologues in HCMV, alluding that some essential genes may perform core functions for all of these herpesviruses.

Another observation worth noting is the size of the essential ORFs as compared to the size of the nonessential ORFs. Essential ORFs are significantly larger in size, averaging 1250-bp, while nonessential ORFs have an average size of 970-bp. Moreover, the ten largest VZV ORFs are all essential, while of the 11 smallest VZV ORFs, eight are nonessential.

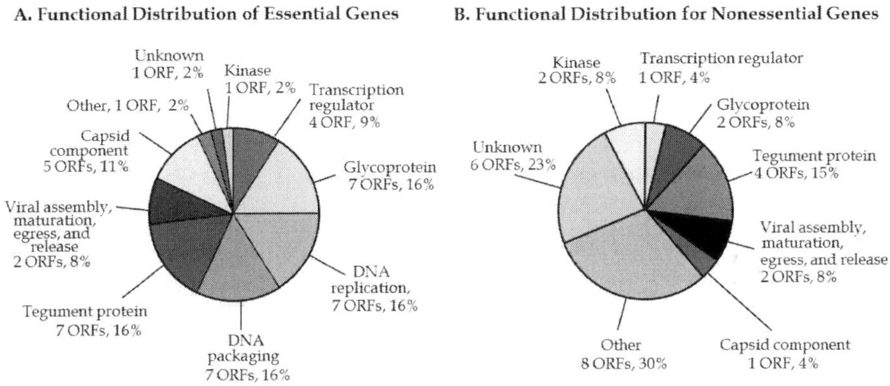

Fig. 9. Distribution of functional annotations for essential and nonessential genes. (A) Distribution of functional annotations for essential genes. Essential genes are significantly enriched for DNA replication (Bonferroni corrected p-value $<10^{-4}$) and DNA packing (corrected p-value $<10^{-4}$) functional categories. (B) Distribution of functional annotations for nonessential genes. Nonessential genes are significantly enriched for other (corrected p-value $<10^{-3}$) and unknown (corrected p-value <0.01) functional categories. Statistical significance was determined by a hypergeometric test.

6. Identification of VZV tissue tropic genes

Although 26 VZV ORFs are shown to be dispensable for viral replication in cultured MeWo cells, it is possible that some of these viral genes may encode proteins critical for optimal viral infection in skin tissue. To test this hypothesis, all of the nonessential ORF deletion mutants were further tested in human fetal skin organ culture (SOC). SOC is a reliable alternative to the SCID-hu mouse model (Taylor & Moffat, 2005), and is especially convenient for initial genome-wide screening of skin-tropism determinants.

We found that all VZV deletions which demonstrated severe growth defects in cultured MeWo cells also exhibited the same growth defects in SOC samples. Interestingly however, among the 18 VZV ORFs believed to be completely dispensable for viral replication in cultured MeWo cells, four ORFs displayed significant growth defects in SOC (Fig. 8) (Zhang et al., 2010). Because these ORFs are trivial for viral replication in MeWo cells but prove crucial for optimal viral replication in skin tissue, they evoked further investigation as potential skin tropism factors.

Rescue viruses were generated for two of these four ORF deletions to ensure that growth defects in skin culture are in fact due to the functions of the deleted genes. As expected, the growth curve analyses showed that in MeWo cells, rescue viruses grew indistinguishably from wild-type VZV, and in SOC, they were able to fully recover the growth defects of their corresponding deletion mutants (Zhang et al., 2010). Three of these four ORFs (ORF10, ORF14, ORF47) have previously been identified as tissue-tropic factors (Cohen & Seidel, 1994; Heineman & Cohen, 1995; Moffat et al., 1998). Our findings verified these previous studies and additionally identified ORF7 as a novel skin-specific virulence factor. To confirm our original finding here, we also produced a premature stop-codon mutant (ORF7S) by mutating the 5th codon from TGT to the TGA stop codon. Like ORF7D, ORF7S displayed wild-type growth in MeWo cells, but had a growth defect in SOC, indicating that ORF7 may function as a VZV skin-tropic factor.

7. Ongoing research

As mentioned previously, after a primary VZV infection, the virus will remain dormant in the sensory ganglia of its host. When reactivated, VZV will erupt from the sensory neurons and infect surrounding skin tissue, causing characteristic rashes and severe pain due to nerve damage. Therefore, identifying the VZV factors responsible for not only skin-tropism, but also neurotropism is of great importance.

7.1 Screening for VZV neurotropic factors

We proposed that the VZV genome also encodes factors required for efficient invasion and egress from specific tissues during natural infection, such as neurotropic factors. Using our newly created VZV deletion mutants, we screened each of the 18 dispensable VZV ORFs to determine which are implicated with VZV neuronal infection. First, the mutant DNAs were confirmed to be replication competent in a primary permissive cell line, the human retinal epithelial ARPE19 cells (Schmidt-Chanasit et al., 2008). Then, the deletion mutants were screened in human neuroblastoma, SH-SY5Y, following a similar transfection approach as described earlier for MeWo cells, to establish their essentiality for replication in human neurons.

Of the mutants tested, only the ORF7 deletion mutant was unable to form viral plaques (data not yet published). To confirm this, we also infected differentiated neuroblastoma and human embryonic stem cell-derived neurons with wild type and 7D cell-free particles. As expected, the WT infection exhibited robust proliferation, while the 7D infection yielded no visible plaques. Our finding here establishes ORF7 as the only known VZV factor required for viral spread in human neurons.

7.2 Future applications

While the current VZV vaccine is sufficient to prevent chickenpox, many issues still surround this live attenuated vaccine. Firstly, bulk vaccine production using the vOka strain is difficult and costly due to vOka virus's relatively low yield (Schmid & Jumaan, 2010; Gomi et al, 2002). The vaccine is also not entirely effective at eliminating chickenpox outbreaks. Despite seroconversion after vaccination, varicella infections still occur in some children and adults exposed to wild-type VZV (Schmid & Jumaan, 2010; Bernstein et al., 1993, White et al., 1991). Most significantly however, the currently marketed vaccine strain, v-Oka, while highly attenuated in the skin, still retains its neurovirulence (Hambleton et al., 2008). This means that the virus will continue to establish latency in the sensory nerve ganglia of the immunized host and can potentially reactivate later to trigger herpes zoster and post-herpetic neuralgia. Furthermore, while drug treatments available to date can alleviate some symptoms of VZV-elicited diseases and shorten the disease duration, they cannot clear the virus or prevent establishment of latency (Miwa et al., 2005; Hatchette et al., 2008). For these reasons, developing a new neuro-attenuated vaccine is imperative to prevent future herpes zoster in both elderly people and vaccinated children.

The use of viral BACs has great potential for novel vaccine development and future treatment of viral diseases. Our studies using the VZV_{Luc} BAC uncovered the first, and plausibly the only, VZV neurotropic factor, ORF7. Because the ORF7D strain is incapable of infecting both human skin and nervous tissue *in vivo*, the deletion virus may serve as an ideal vaccine candidate for the next generation of chickenpox and shingles vaccines. Aside from using the ORF7 deletion virus to produce a safer neuro-attenuated vaccine for the prevention of herpes zoster, the same deletion virus may also be utilized as a potential viral vector for the production of vaccines against other pathogens as well.

Additional research into this BAC-based candidate is needed to lead us closer to designing neuro-attenuated vaccines that do not establish latency in sensory neurons, and thereby eliminate the risk of recurrent herpes zoster and its complications. In time, the development of safe and effective neuro-attenuated vaccines will decrease the likelihood of herpes zoster in the contemporary susceptible population, reduce herpes zoster-associated costs, and potentially eradicate VZV and VZV-related diseases (Lydick et al., 1995; Drolet et al., 2010). Essentially, these new vaccines will change the future of VZV altogether.

8. Conclusions

The use of a bacterial artificial chromosome system has proven to be an invaluable tool in human herpesvirus studies, without which, our genome-wide VZV mutagenesis could not have been possible. Not only can BACs clone the large viral genomic DNA, their slow replication rate and relative ease and accuracy of producing and reproducing stable viral mutants make BACs the ideal method for functional analysis of the VZV genome. Furthermore, the addition of a luciferase marker has greatly improved the efficiency, accuracy and reproducibility of our results. Combined, this luciferase BAC approach has truly facilitated genetic studies of VZV and provided vital insights into the replication and pathogenesis of the virus.

In this study, a global functional analysis of the entire VZV genome was carried out, focusing on the identification of ORFs essential for viral replication in cultured MeWo cells and human fetal skin organs. In all, our study has distinguished novel functional annotations for 36 VZV genes and shed light on the essentiality of each of the 70 unique VZV ORFs. More importantly, our findings have identified ORF7 as both a skin-tropic and neurotropic factor, implicating the ORF7 deletion virus as an ideal vaccine candidate to prevent both VZV-elicited diseases, chickenpox and shingles.

As our research progresses, future VZV_{Luc} BAC studies will continue to provide more exciting discoveries and help identify new antiviral targets. Soon, effective vaccines and improved therapy for the prevention and treatment of a wide array of infections will be tangible.

9. Acknowledgements

We thank Zhen Zhang, Anca Selariu, Charles Warden, Grace Huang, Benjamin Silver, Tong Cheng, Zhongxiang Ye, Ying Huang and Kalpana Dulal for their contributions to the work described in this chapter.

10. References

Abendroth, A. & Arvin, A. (1999). Varicella-zoster virus immune evasion. *Immunol Rev*, Vol. 168, (April 1999), pp. 143-156, ISSN 0105-2896

Arvin, A. M. (1996). Varicella-zoster virus. *Clin Microbiol Rev*, Vol. 9, No. 3, (July 1996), pp. 361-381, ISSN 0893-8512

Arvin, A. M. (2001). Varicella vaccine: genesis, efficacy, and attenuation. *Virology*, Vol. 284, No. 2, (June 2001), pp. 153-158, ISSN 0042-6822

Arvin, A. M. (2006). Investigations of the pathogenesis of varicella zoster virus infection in the SCIDhu mouse model. *Herpes*, Vol. 13, No. 3, (November 2006), pp. 75-80, ISSN 0969-7667

Arvin, A. M., Moffat, J. F., Sommer, M., Oliver, S., Che, X., Vleck, S., Zerboni, L., & Ku, C. C. (2010). Varicella-zoster virus T cell tropism and the pathogenesis of skin infection. *Curr Top Microbiol Immunol*, Vol. 342, (September 2010), pp. 189-209, ISSN 0070-217X

Bernstein, H. H., Rothstein, E. P., Watson, B. M., Reisinger, K. S., Blatter, M. M., Wellman, C. O., Chartrand, S. A., Cho, I., Ngai, A., & White, C. J. (1993). Clinical survey of natural varicella compared with breakthrough varicella after immunization with live attenuated Oka/Merck varicella vaccine. *Pediatrics*, Vol. 92, No. 6, (December 1993), pp. 833-837, ISSN 0031-4005

Besser, J., Sommer, M. H., Zerboni, L., Bagowski, C. P., Ito, H., Moffat, J., Ku, C. C., & Arvin, A. M. (2003). Differentiation of varicella-zoster virus ORF47 protein kinase and IE62 protein binding domains and their contributions to replication in human skin xenografts in the SCID-hu mouse. *J Virol*, Vol. 77, No. 10, (May 2003), pp. 5964-5974, ISSN 0022-538X

Burgoon, C. F., Burgoon, J. S., & Baldridge, G. D. (1957). The natural history of herpes zoster. *J Am Med Assoc*, Vol. 164, No. 3, (May 1957), pp. 265-269

Cohen, J. I. & Seidel K. E. (1993). Generation of Varicella-Zoster Virus (VZV) and Viral Mutants from Cosmid DNAs - VZV Thymidylate Synthetase Is Not Essential for Replication in-Vitro. *Proc Natl Acad Sci USA*, Vol. 90, No. 15, (August 1993), pp. 7376-7380, ISSN 0027-8424

Cohen, J. I. & Seidel, K. (1994). Varicella-zoster virus (VZV) open reading frame 10 protein, the homolog of the essential herpes simplex virus protein VP16, is dispensable for VZV replication in vitro. *J Virol*, Vol. 68, No. 12, (December 1994), pp. 7850-7858, ISSN 0022-538X

Cohen, J. I. & Seidel, K. E. (2001). Varicella-zoster virus and its replication, In: *Fields Virology* (4th ed.), D. M. Knipe & P. M. Howley, pp. 2707-2730, Lippincott Williams & Wilkins, ISBN 0-7818-1832-5, Philadelphia, PA

Cohen, J. , Straus, S., & Arvin, A. (2007). Varicella-Zoster Virus Replication, Pathogenesis, and Management, In: *Fields Virology* (5th ed.), D. M. Knipe & P. M. Howley, pp. 2744-2818, Lippincott Williams & Wilkins, ISBN 0-7817-6060-7, Philadelphia, PA

Contag, C. H., Spilman, S. D., Contag, P. R., Oshiro, M., Eames, B., Dennery, P., Stevenson, D. K., & Benaron, D. A. (1997). Visualizing gene expression in living mammals using a bioluminescent reporter. *Photochem Photobiol*, Vol. 66, No. 4, (October 1997), pp. 523-531, ISSN 0031-8655

Defechereux, P., Melen, L., Baudoux, L., Merville-Louis, M. P., Rentier, B., & Piette, J. (1993). Characterization of the regulatory functions of varicella-zoster virus open reading frame 4 gene product. *J Virol*, Vol. 67, No. 7, (July 1993), pp. 4379-4385, ISSN 0022-538X

Drolet, M., Brisson, M., Schmader, K. E., Levin, M. J., Johnson, R., Oxman, M. N., Patrick, D., Blanchette, C., & Mansi, J. A. (2010). The impact of herpes zoster and postherpetic neuralgia on health-related quality of life: a prospective study. *CMAJ*, Vol. 182, No. 16, (November 2010), pp. 1731-1736, ISSN 1488-2329

Doyle, T. C., Burns, S. M., & Contag, C. H. (2004). In vivo bioluminescence imaging for integrated studies of infection. *Cell Microbiol*, Vol. 6, No. 4, (April 2004), pp. 303-317, ISSN 1462-5814

Dulal, K., Zhang, Z., & Zhu, H. (2009). Development of a gene capture method to rescue a large deletion mutant of human cytomegalovirus. *J Virol Methods*, Vol. 157, No. 2, (May 2009), pp. 180-187, ISSN 0166-0934

Dunn, W., Chou, C., Li, H., Hai, R., Patterson, D., Stolc, V., Zhu, H., & Liu, F. (2003). Functional profiling of a human cytomegalovirus genome. *Proc Natl Acad Sci USA*, Vol. 100, No. 24, (November 2003), pp. 14223-14228, ISSN 0027-8424

Galea, S. A., Sweet, A., Beninger, P., Steinberg, S. P., Larussa, P. S., Gershon, A. A., & Sharrar, R. G. (2008). The safety profile of varicella vaccine: a 10-year review. *J Infect Dis*, Vol. 197, Suppl. 2, (March 2008), S165-169, ISSN 0022-1899

Gershon, A. A. (2001). Live attenuated varicella vaccine. *Infect Dis Clin North Am*, Vol. 15, No. 1, (March 2001), pp. 65-81

Gilden, D. H., Kleinschmidt-DeMasters, B. K., LaGuardia, J. J., Mahalingam, R., & Cohrs, R. J. (2000). Neurologic complications of the reactivation of varicella-zoster virus. *N Engl J Med*, Vol. 342, No. 9, (March 2000), pp. 635-645, ISSN 0028-4793

Gomi, Y., Sunamachi, H., Mori, Y., Nagaike, K., Takahashi, M., & Yamanishi, K. (2002). Comparison of the complete DNA sequences of the Oka varicella vaccine and its parental virus. *J Virol*, Vol. 76, No. 22, (November 2002), pp. 11447-11459, ISSN 0022-538X

Hambleton, S. & Gershon, A. A. (2005). Preventing varicella-zoster disease. *Clin Microbiol Rev*, Vol. 18, No. 1, (January 2005), pp. 70-80, ISSN 0893-8512

Hambleton, S., Steinberg, S. P., Larussa, P. S., Shapiro, E. D., Gershon, A. A. (2008). Risk of herpes zoster in adults immunized with varicella vaccine. *J Infect Dis*, Vol. 197, Suppl. 2, (March 2008), S196-199, ISSN 0022-1899

Hastings, J. W. (1983). Biological diversity, chemical mechanisms, and the evolutionary origins of bioluminescent systems. *J Mol Evol*, Vol. 19, No. 5, (January 1983), pp. 309-321, ISSN 0022-2844

Hatchette, T., Tipples, G. A., Peters, G., Alsuwaidi, A., Zhou, J., & Mailman, T. L. (2008). Foscarnet salvage therapy for acyclovir-resistant varicella zoster: report of a novel thymidine kinase mutation and review of the literature. *Pediatr Infect Dis J*, Vol. 27, No. 1, (January 2008), pp. 75-77, ISSN 0891-3668

Heineman, T. C. & Cohen, J. I. (1994). Deletion of the varicella-zoster virus large subunit of ribonucleotide reductase impairs growth of virus in vitro. *J Virol*, Vol. 68, No. 5, (May 1994), pp. 3317-3323, ISSN 0022-538X

Heineman, T. C. & Cohen, J. I. (1995). The varicella-zoster virus (VZV) open reading frame 47 (ORF47) protein kinase is dispensable for viral replication and is not required for phosphorylation of ORF63 protein, the VZV homolog of herpes simplex virus ICP22. *J Virol*, Vol. 69, No. 11, (November 1995), pp. 7367-7370, ISSN 0022-538X

Ku, C. C., Besser, J., Abendroth, A., Grose, C., & Arvin, A. M. (2005). Varicella-zoster virus pathogenesis and immunobiology: new concepts emerging from investigations with the SCIDhu mouse model. *J Virol*, Vol. 79, No. 5, (March 2005), pp. 2651-2658, ISSN 0022-538X

Kurfurst, M., Ghisla, S., & Hastings, J. W. (1983). Bioluminescence emission from the reaction of luciferase-flavin mononucleotide radical with O2. *Biochemistry*, Vol. 22, No. 7, (March 1983), pp. 1521-1525, ISSN 0006-2960

Lydick, E., Epstein, R. S., Himmelberger, D., & White, C. J. (1995). Herpes zoster and quality of life: a self-limited disease with severe impact. *Neurology*, Vol. 45, No. 12, Suppl. 8, (December 1995), S52-53, ISSN 0028-3878

Mallory, S., Sommer, M., & Arvin, A. M. (1997). Mutational analysis of the role of glycoprotein I in varicella-zoster virus replication and its effects on glycoprotein E conformation and trafficking. *J Virol*, Vol. 71, No. 11, (November 1997), pp. 8279-8288, ISSN 0022-538X

Mallory, S., Sommer, M., & Arvin, A. M. (1998). Analysis of the glycoproteins I and E of varicella-zoster virus (VZV) using deletional mutations of VZV cosmids. *J Infect Dis*, Vol. 178, Suppl. 1, (November 1998), S22-26, ISSN 0022-1899

Marchini, A., Liu, H., & Zhu, H. (2001). Human Cytomegalovirus with IE-2 (UL122) Deleted Fails To Express Early Lytic Genes. *J. Virol*, Vol. 75, No. 4, (February 2001), pp. 1870-1878, ISSN 0022-538X

Miwa, N., Kurosaki, K., Yoshida, Y., Kurokawa, M., Saito, S., & Shiraki, K. (2005). Comparative efficacy of acyclovir and vidarabine on the replication of varicella-zoster virus. *Antiviral Res*, Vol. 65, No. 1, (January 2005), pp. 49-55, ISSN 0166-3542

Moffat, J. F., Zerboni, L., Kinchington, P. R., Grose, C., Kaneshima, H., & Arvin, A. M. (1998). Attenuation of the vaccine Oka strain of varicella-zoster virus and role of glycoprotein C in alphaherpesvirus virulence demonstrated in the SCID-hu mouse. *J Virol*, Vol. 72, No. 2, (February 1998), pp. 965-974, ISSN 0022-538X

Moffat, J. F. & Arvin, A. M. (1999). Varicella-zoster virus infection of T cells and skin in the SCID-hu mouse model, In *Handbook of animal models of infection: experimental models in antimicrobial chemotherapy*, M. A. Sande & O. Zak, pp. 973-980, Academic Press, ISBN 978-0-12-775390-4, San Diego, CA

Moriuchi, H., Moriuchi, M., Straus, S. E., & Cohen, J. I. (1993). Varicella-zoster virus (VZV) open reading frame 61 protein transactivates VZV gene promoters and enhances the infectivity of VZV DNA. *J Virol*, Vol. 67, No. 7, (July 1993), pp. 4290-4295, ISSN 0022-538X

Moriuchi, H., Moriuchi, M., Smith, H. A., & Cohen, J. I. (1994). Varicella-zoster virus open reading frame 4 protein is functionally distinct from and does not complement its herpes simplex virus type 1 homolog, ICP27. *J Virol*, Vol. 68, No. 3, (March 1994), pp. 1987-1992, ISSN 0022-538X

Nagaike, K., Mori, Y., Gomi, Y., Yoshii, H., Takahashi, M., Wagner, M., Koszinowski, U., & Yamanishi, K. (2004). Cloning of the varicella-zoster virus genome as an infectious bacterial artificial chromosome in Escherichia coli. *Vaccine*, Vol. 22, No. 29-30, (September 2004), pp. 4069-4074, ISSN 0264-410X

Niizuma, T., Zerboni, L., Sommer, M. H., Ito, H., Hinchliffe, S., & Arvin, A. M. (2003). Construction of Varicella-Zoster Virus Recombinants from Parent Oka Cosmids and Demonstration that ORF65 Protein Is Dispensable for Infection of Human Skin and T Cells in the SCID-hu Mouse Model. *J. Virol*, Vol. 77, No. 10, (May 2003), pp. 6062-6065, ISSN 0022-538X

Opstelten, W., McElhaney, J., Weinberger, B., Oaklander, A. L., & Johnson, R. W. (2010). The impact of varicella zoster virus: chronic pain. *J Clin Virol*, Vol. 48, Suppl. 1, (May 2010), S8-13 , ISSN 1873-5967

Oxman, M. N., Levin, M. J., Johnson, G. R., Schmader, K. E., Straus, S. E., Gelb, L. D., Arbeit, R. D., Simberkoff, M. S., Gershon, A. A., Davis, L. E., Weinberg, A., Boardman, K. D., Williams, H. M., Zhang, J. H., Peduzzi, P. N., Beisel, C. E., Morrison, V. A., Guatelli, J. C., Brooks, P. A., Kauffman, C. A., Pachucki, C. T., Neuzil, K. M., Betts, R. F., Wright, P. F., Griffin, M. R., Brunell, P., Soto, N. E., Marques, A. R., Keay, S. K., Goodman, R. P., Cotton, D. J., Gnann, J. W., Jr., Loutit, J., Holodniy, M., Keitel, W. A., Crawford, G. E., Yeh, S. S., Lobo, Z., Toney, J. F., Greenberg, R. N., Keller, P. M., Harbecke, R., Hayward, A. R., Irwin, M. R., Kyriakides, T. C., Chan, C. Y., Chan, I. S., Wang, W. W., Annunziato, P. W., Silber, J. L. (2005). A vaccine to prevent herpes zoster and postherpetic neuralgia in older adults. *N Engl J Med*. Vol. 352, No. 22, (June 2005), pp. 2271-2284, ISSN 1533-4406

Perera, L. P., Mosca, J. D., Sadeghi-Zadeh, M., Ruyechan, W. T., & Hay, J. (1992). The varicella-zoster virus immediate early protein, IE62, can positively regulate its cognate promoter. *Virology*, Vol. 191, No. 1, (November 1992), pp. 346-354, ISSN 0042-6822

Pickering, G. & Leplege, A. (2010). Herpes zoster pain, postherpetic neuralgia, and quality of life in the elderly. *Pain Pract*, Vol. 11, No. 4, (July-August 2011), pp. 397-402, ISSN 1533-2500

Reddy, S. M., Cox, E., Iofin, I., Soong, W., & Cohen, J. I. (1998). Varicella-zoster virus (VZV) ORF32 encodes a phosphoprotein that is posttranslationally modified by the VZV ORF47 protein kinase. *J Virol*, Vol. 72, No. 10, (October 1998), pp. 8083-8088, ISSN 0022-538X

Rehemtulla, A., Stegman, L. D., Cardozo, S. J., Gupta, S., Hall, D. E., Contag, C. H., & Ross, B. D. (2000). Rapid and quantitative assessment of cancer treatment response using in vivo bioluminescence imaging. *Neoplasia*, Vol. 2, No. 6, (November-December 2000), pp. 491-495, ISSN 1522-8002

Rehemtulla, A., Hall, D. E., Stegman, L. D., Prasad, D., Chen, G., Bhojani, M. S., Chenevert, T. L., & Ross, B. D. (2002). Molecular imaging of gene expression and efficacy following adenoviral-mediated brain tumor gene therapy. *Mol Imaging*, Vol. 1, No. 1, (January-March 2002), pp. 43-55, ISSN 1535-3508

Roizman, B., Knipe, D. M., & Whitley, R. J. (2007). Herpes simplex virus, In: *Fields Virology* (5th ed.), D. M. Knipe & P. M. Howley, pp. 2399-2460, Lippincott Williams & Wilkins, ISBN 0-7817-6060-7, Philadelphia, PA

Schmid, D. S. & Jumaan, A. O. (2010). Impact of varicella vaccine on varicella-zoster virus dynamics. *Clin Microbiol Rev*, Vol. 23, No. 1, (January 2010), pp. 202-217, ISSN 1098-6618

Schmidt-Chanasit, J., Bleymehl, K., Rabenau, H. F., Ulrich, R. G., Cinatl, J. Jr., & Doerr, H. W. (2008). In vitro replication of varicella-zoster virus in human retinal pigment epithelial cells. *J Clin Microbiol*, Vol. 46, No. 6, (June 2008), pp. 2122-2124, ISSN 1098-660X

Takahashi, M., Otsuka, T., Okuno, Y., Asano, Y., & Yazaki, T. (1974). Live vaccine used to prevent the spread of varicella in children in hospital. *Lancet*, Vol. 2, No. 7892, (November 1974), pp. 1288-1290, ISSN 0140-6736

Tang, Q., Zhang, Z., & Zhu, H. (2008). Bioluminescence Imaging for Herpesvirus Studies in vivo, In: *Herpesviridae: Viral Structure, Life Cycle and Infections*, T. R. Gluckman, Nova Science Publishers, Inc., ISBN 978-1-60692-947-6

Taylor, S. L. & Moffat, J. F. (2005). Replication of varicella-zoster virus in human skin organ culture. *J Virol*, Vol. 79, No. 17, (September 2005), pp. 11501-11506, ISSN 0022-538X

Volpi, A. (2005). Varicella immunization and herpes zoster. *Herpes*, Vol. 12, No. 3, (December 2005), p. 59, ISSN 0969-7667

Warden, C., Tang, Q., & Zhu, H. (2010). Herpesvirus BACs: past, present, and future. *J Biomed Biotechnol*, Vol. 2011, (October 2010), 124595, ISSN 1110-7251

White, C. J., Kuter, B. J., Hildebrand, C. S., Isganitis, K. L., Matthews, H., Miller, W. J., Provost, P. J., Ellis, R. W., Gerety, R. J., & Calandra, G. B. (1991). Varicella vaccine (VARIVAX) in healthy children and adolescents: results from clinical trials, 1987-1989. *Pediatrics*, Vol. 87, No. 5, (May 1991), pp. 604-610, ISSN 0031-4005

Yamagishi, Y., Sadaoka, T., Yoshii, H., Somboonthum, P., Imazawa, T., Nagaike, K., Ozono, K., Yamanishi, K., & Mori, Y. (2008). Varicella-zoster virus glycoprotein M homolog is glycosylated, is expressed on the viral envelope, and functions in virus cell-to-cell spread. *J Virol*, Vol. 82, No. 2, (January 2008), pp. 795-804, ISSN 1098-5514

Yu, D. G., Ellis, H. M., Lee, E. C., Jenkins, N. A., Copeland, N. G., & Court, D. L. (2000). An efficient recombination system for chromosome engineering in Escherichia coli. *Proc Natl Acad Sci USA*, Vol. 97, No. 11, (May 2000), pp. 5978-5983, ISSN 0027-8424

Zerboni, L., Ku, C. C., Jones C. D., Zehnder, J. L., & Arvin, A. M. (2005). Varicella-zoster virus infection of human dorsal root ganglia in vivo. *Proc Natl Acad Sci USA*, Vol. 102, No. 18, (May 2005), pp. 6490-6495, ISSN 0027-8424

Zerboni, L., Reichelt, M., & Arvin, A. (2010). Varicella-zoster virus neurotropism in SCID mouse-human dorsal root ganglia xenografts. *Curr Top Microbiol Immunol*, Vol. 343, (March 2010), pp. 255-276, ISSN 0070-217X

Zhang, Z., Rowe, J, Wang, W., Sommer, M., Arvin, A., Moffat, J., & Zhu, H. (2007). Genetic analysis of varicella-zoster virus ORF0 to ORF4 by use of a novel luciferase bacterial artificial chromosome system. *J Virol*, Vol. 81, No. 17, (September 2007), pp. 9024-9033, ISSN 0022-538X

Zhang, Z., Huang, Y., & Zhu, H. (2008). A highly efficient protocol of generating and analyzing VZV ORF deletion mutants based on a newly developed luciferase VZV BAC system. *Journal of Virological Methods*, Vol. 148, No. 1-2, (March 2008), pp. 197-204, ISSN 0166-0934

Zhang, Z., Selariu, A., Warden, C., Huang, G., Huang, Y, Zaccheus, O., Cheng, T., Xia, N., & Zhu, H. (2010). Genome-Wide Mutagenesis Reveals That ORF7 Is a Novel VZV Skin-Tropic Factor. *PLoS Pathogens*, Vol. 6, No. 7, (July 2010), e1000971

6

Gene Functional Studies Using Bacterial Artificial Chromosome (BACs)

Mingli Liu, Shanchun Guo, Monica Battle and Jonathan K. Stiles
Microbiology, Biochemistry and Immunology
Morehouse School of Medicine, Atlanta
USA

1. Introduction

1.1 Background and history of bacterial artificial chromosome (BACs)

BACs were first developed as a large insert cloning system to facilitate the construction of DNA libraries to analyze genomic structure (Shizuya, Birren et al. 1992). BACs are derived from a fertility plasmid (F-plasmid) found in the *Escherichia coli.* BACs can clone extremely large DNA molecules, ranging from 150-700kb,and averaging 350 kb. Another advantage of BACs over other cloning technologies is its stability in cell culture and ease of manipulation. Because some recombinant viruses were too large to be generated by traditional techniques, BAC technology was developed to carry out genetic and functional studies of viruses, especially herpesvirus. (Warden, Tang et al. 2011). From the time when BACs emerged a decade ago, their application have grown intensely and have benefited the research community in many fields, such as sequencing of the human genome, *in vitro* transgenesis, genomic fingerprinting, and even to vaccine development.

1.2 BACs generation

BACs can be considered as plasmid expression vectors, amplified in bacteria and composed of a small amount of bacterial DNA derived from the single copy F-plasmid (Shizuya, Birren et al. 1992). F-plasmid has genes required for prokaryotic replication, partition, and selection. Viral BACs are generated by inserting a BAC vector sequence into a viral genome (Warden, Tang et al. 2011). Direct deletion mutants and random transposon mutagenesis of viral BACs are commonly used to determine the function of viral genes (Warden, Tang et al. 2011). Mammalian DNA, human genome or mouse genome (100–350 kb) can also be cloned as BACs vectors. Currently most regions of the human genome and the genome of other species are available as BACs. These vectors are useful tools in human genome-sequencing projects (Lander, Linton et al. 2001; Venter, Adams et al. 2001) and transgenic mice studies. Using endogenous homologous recombination systems in *E. coli* (Copeland, Jenkins et al. 2001), BACs can be modified in many ways by recombination. They can be genetically engineered to express reporter genes, or to put a transgene under a specific promoter which will be more amenable in gene expression in the mammalian cells. BACs can be modified to replace any nucleotide sequence of interest (gene replacement), remove any existing DNA sequence, or introduce new sequences without removing any of the existing sequences (gene removal or insertion). In addition BACs are also used to place nucleotide substitution through selection/counterselection strategy, and to conduct effective gap repair cloning of any target site of interest (Adamson, Jackson et al. 2011).

1.3 Advantages and disadvantages of BACs

Genes expressed from BACs mirror endogenous gene expression far more accurately than other cloning systems. The large size of BACs help to minimize site of integration effects, a phenomenon which has been defined as endogenous sequences (such as gene coding regions and distal regulatory elements) to be disrupted, and to produce potentially undesirable phenotypes (Adamson, Jackson et al. 2011) in gene cloning technology. The larger sized BAC constructs contain enhancers and locus control regions, which leads to more accurate gene expression *in vivo* (Townes, Lingrel et al. 1985; Jones, Monks et al. 1995). 1995). The human genome BACs consist of the full gene structure, including untranslated regions, exons and introns, alternative promoters and splice sites and microRNA coding sequences. RNAs such as RNA splicing or microRNAs play very important role in gene regulation (Jackson and Standard 2007). Therefore the human genome BACs will ensure full mRNA processing and splicing when genes are transcribed, and produce the full complement of protein isoforms once mRNAs are translated. BACs can be transfected and expressed in mammalian cell lines although transfection efficiency and copy numbers are low (Magin-Lachmann, Kotzamanis et al. 2004; Sparwasser and Eberl 2007).

BACs also have a number of disadvantages. A construct containing a large genomic fragment is likely to contain non-related genes that may lead to indirect, non-specific gene expression and unanticipated changes in the cell phenotype; Secondly, compared to plasmids or other gene expression vectors, the generation and screening of recombinant BAC constructs can be time-consuming and labor-intensive. Also, the oversized BAC DNA constructs are more easily sheared and degraded during manipulation before transfection; and some random recombination events may occur, for example, LoxP sites may lead to random Cre-mediated recombination (Semprini, Troup et al. 2007). Finally, repeating homologous sequences in some BACs constructs may undergo intramolecular rearrangements, which reduce the recombination efficiency and increase the rate of false-positive clones in some selection/counter-selection approaches (Narayanan 2008).

Overall, BACs have numerous advantages when compared to conventional plasmids. They protect the gene from site of integration effects and produce accurate regulation of transcription and translation. However, the large size results in technical difficulties when handling them as well as the potential non-specific gene expression. Therefore the application of BACs as a gene expression model system should be careful considered based on the pros and cons previously described.

1.4 Application of BACs: Genomic sequencing, genomic imprinting, transgenic mice, vaccine development, and gene therapy

There is increasing interest in the application of BAC technology in genomic research. High throughput determination of gains and losses of genetic material using high resolution BAC arrays and comparative genomic hybridization (CGH) have been developed into the new tools for translational research in solid tumors and neurodegenerative disorders (Cowell and Nowak 2003; Cowell 2004; Costa, Meijer et al. 2008; Lu 2009). Among a large number of approaches for sequencing, BAC technology is becoming the most robust method for genome sequencing. The BAC-by-BAC technique uses an overlapping tilling path of large genomic fragments (150-200 kb) maintained within BACs. Every individual BAC is shotgun sequenced. Many short reads are assembled to produce the sequence of the BACs, where these large overlapping sequences of the BACs are assembled to produce the who genome sequence (Imelfort, Batley et al. 2009). BACs have also been used in mammalian genome mapping (Schalkwyk, Francis et al. 1995), genomic imprinting (Tunster, Van De Pette et al. 2011), vaccine development and gene therapy (Magin-Lachmann, Kotzamanis et al. 2004; Warden, Tang et al. 2011). Studies of the evolutionary history and functional dynamics of sex chromosomes have recently been possible using BAC libraries (Janes, Valenzuela et al. 2011). In this chapter we will review some applications of BACs in viral and non-viral gene functional studies.

2. Viral gene functional studies

2.1 Many human and animal herpesviruses genomes have been cloned as BACs

Human herpesviruses are the second leading cause of human viral disease. Therefore the utilization of human herpesvirus BACs to study viral gene function (Warden, Tang et al. 2011) has become more and more common. The herpesviruses are a family of DNA viruses which contain large and complex genomes. Genetic control and management of recombinant viruses have been notoriously difficult. The development of herpesvirus BACs have facilitated generation of recombinant viruses and subsequent studies of the biology and pathogenesis of herpesviruses (Knipe, Batterson et al. 1981; Zhou and Roizman 2005). Table 1 shows the human herpesviruses which have been cloned as BACs, including Herpes simples virus type 1 [(HSV-1 or human herpesvirus (HHV-1)], varicella-zoster virus (VZV or HHV-3), human cytomegalovirus (HCMV or HHV-5), Kaposi's sarcoma-associated herpesvirus (KSHV or HHV-8) (Feederle, Bartlett et al. 2010; Warden, Tang et al. 2011). In general, BAC clones are relatively easy to make for alpha- and beta herpes viruses than gamma herpes viruses. This is due to the fact that DNA can only persistently stay in bacterial cells when it has a prokaryotic replicon. When a BACs flanked by specific Herpesvirus genomic sequences were introduced into infected cells to trigger homologous recombination. The great efficiency was achieved in alpha- and beta herpes viruses because lytic cellular systems are available, but was difficult for gamma herpesviruses (Delecluse, Hilsendegen et al. 1998; Delecluse, Kost et al. 2001; Zhou, Zhang et al. 2002; Kanda, Yajima et al. 2004; Chen, Li et al. 2007). In addition to BAC-based human herpesvirus studies, BAC-based other animal herpesviruses are also currently available. These include murine cytomegalovirus 68 (MHV-68), murine gammaherpesvirus (mCMV), rhesus cytomegalovirus (rhCMV), rhesus rhadinovirus (RRV), pseudorabies virus (PrV), herpesvirus saimiri (HVS), Marek's disease virus (MDV), bovine herpesvirus type 1 (BHV-1), equine herpesvirus type 1 (EHV-1), feline herpesvirus (FHV-1), guinea pig cytomegalovirus (GPCMV), Koi herpesvirus (KHV) and turkery herpesvirus (HVT) (Feederle, Bartlett et al. 2010; Warden, Tang et al. 2011).

type	synonym	subfamily	biological function and application	reference
HHV-1	Herpes simplex virus-1 (HSV-1)	α (Alpha)	Generates a replication-proficient but packaging-deficient HSV-1 genome (152-kb HSV-1) for genetic manipulation as research tools or vectors in gene therapy.	(Saeki, Ichikawa et al. 1998; Stavropoulos and Strathdee 1998; Horsburgh, Hubinette et al. 1999)
HHV-2	Herpes simplex virus-2 (HSV-2)	α	Generates a recombinant HSV-2 BAC with the deletion of the HSV-2 glycoprotein D (gD), elicites an HSV-2 specific antibody response, serves as the basis for novel HSV-2 vaccine production.	(Meseda, Schmeisser et al. 2004)
HHV-3 Varicella zoster virus (VZV)		α	Human embryonic lung cells transfected with VZV BAC DNA show cytopathic effect, and viruses can spread to neighboring cells.	(Nagaike, Mori et al. 2004)
			Luciferase VZV BAC generates recombinant VZV variants, eases subsequent viral growth kinetic analysis both *in vitro* MeWo cells and SCID-hu mice *in vivo*.	(Zhang, Rowe et al. 2007)
			The mini-F transposition technique optimizes, repairs or restructures BACs, facilitates the development of gene therapy or vaccine vectors.	(Wussow, Fickenscher et al. 2009)
HHV-4 Epstein-Barr virus (EBV), lymphoc ryptovirus (LCV)		γ (Gamma)	Genetic analysis of all EBV functions, generation of attenuated EBV strains for vaccine design, development of viral vectors for human gene therapy.	(Delecluse, Hilsendegen et al. 1998; Kanda, Yajima et al. 2004)
			Generates a self-recombining BAC containing 172-kb of the EBV genome; provides proof that EBNA-3B is not essential for EBV-mediated B-cell growth transformation *in vitro*.	(Chen, Divisconte et al. 2005)

type	synonym	subfamily	biological function and application	reference
HHV-5 Cytomegalovirus (HCMV)		β (Beta)	A new approach for construction of HCMV mutants.	(Borst, Hahn et al. 1999)
			IE-2 (UL122) is required for successful HCMV infection and indicates that virus lacking IE-2 arrests early in the replication cycle.	(Marchini, Liu et al. 2001)
			UL45 is dispensable for growth of HCMV in human fibroblasts and human endothelial cells	(Hahn, Khan et al. 2002)
			A total of 252 ORFs with the potential to encode proteins have been identified in two laboratory strains (AD169 and Towne) and four clinical isolates Toledo, FIX, PH, and TR	(Murphy, Yu et al. 2003)
			HCMV strain TB40/E is available as a BAC clone suitable for genetic engineering.	(Sinzger, Hahn et al. 2008)
			Develops a "gene capture" method to rescue a large deletion mutant (15kb) of HCMV.	(Dulal, Zhang et al. 2009)
HHV-6	Roseolovirus, Herpes lymphotropic virus	β	Develops a single-step production of viral BACs by introducing the 160-kb human herpes virus 6A genome into BACs by digesting the viral DNA replicative intermediates with the SfiI enzyme that cleaves the viral genome in a single site	(Borenstein and Frenkel 2009)
HHV-7	Roseolovirus	β		Not available
HHV-8	KSHV	γ	BAC KSHV can be efficiently shuttled between bacteria and mammalian cells, such as BCBL-1, 293, HeLa and E6E7-immortalized human endothelial cells.	(Delecluse, Kost et al. 2001; Zhou, Zhang et al. 2002; Liu 2010)

Table 1. List of available BACs for human HSV

2.2 vGPCR-mediated angiogenesis through activation of p38 and STAT3 in KSHV infected cells using KSHV BACs

The molecular mechanism whereby viral G protein-coupled receptor (vGPCR) signaling regulates vascular endothelial growth factor (VEGF) expression in Kaposi sarcoma (KS) formation remains somewhat undefined. mECK36 cells, generated by transfection of mice bone marrow endothelial cells with KSHV bacterial artificial chromosome (KSHVBac36), have been reported to be angiogenic, tumorigenic, and suitable for demonstrating a nonredundant role for vGPCR in KSHV-mediated tumorigenesis (Mutlu, Cavallin et al. 2007). In our previous report (Liu 2010), we utilized mECK36, the cells composed of wild-type KSHVBac36 or the cells without vGPCR, namely vGPCR-null KSHVBac36 mutant, to dissect the molecular mechanisms of VEGF secretion induced by vGPCR in the context of KSHV infection. The mice bone marrow endothelial cells (mEC) were obtained from Balb/C An Ncr-nu mice (NCI, Bethesda, MD) bone marrow. Mice femurs were flushed twice with phosphate-buffered saline (PBS), and the elutes were incubated in Dulbecco's modified Eagle's medium (DMEM) media plus 30% fetal borine serum (FBS) (Gemini Bioproducts, Calabasas, CA), endothelial growth factor (EGF) 0.2 mg/mL (Sigma, St. Louis, MO), endothelial cell growth factor supplement (ECGS) 0.2 mg/mL (Sigma), heparin 1.2 mg/L (Sigma), insulin transferrine selenium (Invitrogen, Carlsbad, CA), penicilin-streptomicin 1% (Invitrogen), and BME vitamin (VWR Scientific, Rochester, NY). KSHVBac36 was constructed by inserting a full-length recombinant KSHV genome into a bacterial artificial chromosome, KSHVBac36 was transfected into mEC cells to generate mECK36 cells using lipofectamine 2000 (Invitrogen) and selected with hygromycin-B. The cells were then grown in the absence of hygromycin to negatively select cells and therefore generate mECK36-KSHV-Null cells, which lost the KSHV episome (KSHV episome was measured by GFP marker). Next, KSHVBac36 construct was retransfected into mECK36-KSHV-Null cells to generate BBac36. Finally, the genotypic markers of vGPCR were knocked out from KSHVBac36 by transposon mutagenesis to generate ORF74/vGPCR deletion mutant and stably transfected into mECK36-KSHV-Null cells to create BΔvGPCR cells in the presence of hygromycin selection. We found (Liu 2010) that vGPCR activates VEGF transcription via p38 MAPK and STAT3 in mECK36 and mECK36-derived cell models. In addition, we also found that in cells containing KSHV genome, STAT3 is tyrosine-phosphorylated and translocated into the nucleus, transactivating the target VEGF gene by binding to the specific DNA element TT (N4–5) AA in a vGPCR-dependent manner. Moreover, treatment of mECK36-derived cells with AG490 or a dominant negative STAT3 DNA vector showed strong inhibitory effects on vGPCR-induced VEGF promoter activity. In addition, vGPCR can up-regulate STAT3 mRNA levels. Together, our findings show that vGPCR plays a nonredundant role in STAT3 activation in KSHV infected cells, and this activation plays an important role in the connection of the viral oncogene vGPCR and VEGF up-regulation. Our results indicate that vGPCR has a broad signaling activating capacity in the context of KSHV infection and suggest that the STAT3 pathway could be a good target for preventing KSHV-mediated angiogenesis in KS.

2.3 Genetic determinants of virus tropism genes using BACs

Many cell types, including endothelial cells (ECs), myeloid lineage cells, and smooth muscle cells are permissive cells for HCMV persistent replication and latency (Jarvis and Nelson 2007). During acute infection of CMV in immune-compromised patients, a number of cell types, such as ECs, various leukocytes, epithelial cells, hepatocytes, smooth muscle cells, and fibroblasts, can be infected because of uncontrolled replication of viruses (Howell, Miller et al. 1979; Myerson, Hackman et al. 1984; Gnann, Ahlmen et al. 1988; Wiley and Nelson 1988; Dankner, McCutchan et al. 1990; Sinzger, Grefte et al. 1995; Read, Zhang et al. 1999; Bissinger, Sinzger et al. 2002). ECs appear to play a critical role in the process of HCMV persistent active infection and maintenance within the host, which is controlled by genetic determinants. Previous studies observed that HCMV strains differed in their ability to infect ECs, which are called EC tropism (MacCormac and Grundy 1999; Sinzger, Schmidt et al. 1999; Kahl, Siegel-Axel et al. 2000). The research on EC tropism has been strengthened by the availability of genetically stable CMV BACs and subsequent mutagenesis of these BACs (Brune, Menard et al. 2001; Scrivano, Sinzger et al. 2011). The switch of cell tropism in different cell types after alternate replication might direct infection from one cell type to the other.

The typical model for tropism is the difference in cell tropism of virus released from EC and fibroblasts. Supernatants from infected human foreskin fibroblasts (HFF) showed a higher ability to infect EC than EC-derived supernatants (Scrivano, Sinzger et al. 2011). Scrivano et al (Scrivano, Sinzger et al. 2011) using mutagenesis of the BAC-cloned HCMV strain TB40/E (TB40-BAC4) found that ECs release a virus progeny of unEC-tropic (not EC-tropic), and retain a progeny of highly EC-tropic; while HFF release both EC-tropic and non EC-tropic virus progeny, HFF progeny is composed of both EC-tropic and non EC-tropic virus populations. The biochemical basis for this phenomenon is due to a different level of gH/gL/pUL(128,130,131A) complex in virions (Scrivano, Sinzger et al. 2011). The CMV EC tropism has been characterized by a "genomic tropism island" composed of three open reading frames (ORFs): UL128, UL130, and UL131A. The region of these genes is important for EC tropism (Hahn, Revello et al. 2004; Scrivano, Sinzger et al. 2011). EC-tropic population most likely is a population with a high gH/gL/pUL(128,130,131A) content. UL128, UL130, and UL131A are required for replication of HCMV in HUVECs (Hahn, Revello et al. 2004). EC tropism for HCMV is highly dependent on the roles of pUL128, pUL130, and pUL131A (Jarvis and Nelson 2007) in virions. EC-tropism produced by an EC-tropic progeny released by HFF, can be depleted with antibodies directed against pUL131A. They propose that the difference in cell tropism of virus released from EC and fibroblasts is caused by a sorting process. EC strongly and specifically retain EC-tropic viruses through the gH/gL/pUL(128,130,131A) complex. Thus, the levels of gH/gL/pUL(128,130,131A) complexes could define whether a particle is EC-tropic or not. A disulfide-linked complex between gH/gL glycoproteins is required for viral entry and fusion. The gH/gL exists in two distinct forms, one composed of pUL128, pUL130, and pUL131A. The pUL128 and pUL130 proteins are linked with gH/gL; pUL131A is required for infection of ECs. The second distinct form is composed of gO alone; the gO protein is linked with gH/gL and is required for replication in fibroblasts. The gH/gL/pUL128/pUL130/pUL131A unit in virions is mandatory for access into ECs which are pH-dependent. Whereas the gH/gL/gO unit in virions are mandatory for access into fibroblasts which are also pH-independent (Jarvis and Nelson 2007). Recently, results from Wang et al showed (Wang, Yu et al. 2007) that HCMV progenies derived from epithelial cells and fibroblasts are also different. It seems the propensity of cells to release viruses plays a crucial role in the establishment of infection and transfer of viruses to new hosts or the fetus.

In addition to HCMV, EBV also works as a cell type-tropic virus. Hutt-Fletcher et al (Hutt-Fletcher 2007) has established the paradigm that epithelial cells produce a EBV virus progeny with high levels of gH/gL/gp42 complexes, facilitating B-cell infection. B-cells in turn, generate virus progeny with low levels of gH/gL/gp42 complexes which efficiently infect epithelial cells, but not B cells. To some extent, this relative switch of cell tropism after alternate replication in epithelial and B-cells directs infection from one cell type to the other.

2.4 Study of the immune response against the EBV using EBV BACs

There are three recombinant wild-type EBVs that have been generated so far (Delecluse, Hilsendegen et al. 1998; Kanda, Yajima et al. 2004; Chen, Divisconte et al. 2005). They were generated by the insertion of the prokaryotic F-plasmid (F-factor) in two B95.8 or one Akata strains. Although the insertion sites differ in these three EBV BACs, at the site of the B95.8 deletion (Delecluse, Hilsendegen et al. 1998), or at the major internal repeat region of the B95.5 strain (Chen, Divisconte et al. 2005), or at BXLF1 open reading frame (ORF) in Akata strain (Kanda, Yajima et al. 2004), the insertion site of the F-plasmid does not affect the phenotype of the virus.

2.4.1 EBV infection

EBV is tightly related to the development of many human cancers. Chen et al (Chen, Divisconte et al. 2005) has developed a BAC-GFP-EBV (containing 172-kb of the EBV genome) system to monitor early cellular and viral events associated with EBV infection. BAC-GFP-EBV was transfected into the HEK 293T epithelial cell line (Halder, Murakami et al. 2009). Then the progeny virus produced by a chemical was used to immortalize human primary B-cell which can be easily monitored by green fluorescence and proliferation. The results showed a dramatic increase in Ki-67, CD40, and CD23 signals. The viral genes express a pattern of an early burst of lytic gene expression. This up-regulation of lytic gene expression prior to latent genes during early infection strongly suggests that the resulting progeny virus is capable of infecting new primary B-cells (Halder, Murakami et al. 2009). This process may be critical for establishment of latency prior to cellular transformation (Halder, Murakami et al. 2009).

2.4.2 EBV transformation

EBV is associated with a number of human malignancies. There is increasing research interest in the molecular functions of these EBV gene products in transformation and evasion from host immune surveillance systems (Izumi 2001). BAC technology made the study on the molecular function of EBV transforming genes feasible because some latent genes such as EBNA1 cannot be maintained in latently infected B cells using traditional cosmid technology (Izumi 2001; Feederle, Bartlett et al. 2010). EBNA1 was found to function as a transactivator of other latent proteins, and was required for replication of the viral genome (Altmann, Pich et al. 2006). When 71kb of EBV DNA genome was amplified in *E.coli* and transfected into primary B-lymphocyes, Altmann et al (Altmann, Pich et al. 2006) identified that EBV DNA is sufficient to immortalize primary human B lymphocytes. Kempkes et al (Kempkes, Pich et al. 1995; Izumi 2001) also identified EBNA3a as a transforming gene, which contributes primarily to the initiation of cell proliferation (Kempkes, Pich et al. 1995; Izumi 2001). Two genes BALF1 and BHRF1 which encode homologous cellular antiapoptotic viral Bcl-2 proteins (vBcl-2), were suggested to interfere with the cell apoptosis program to counteract cell death, which protects the virus from apoptosis in its host cell during virus synthesis (Altmann and Hammerschmidt 2005).

2.4.3 Immune evasion

Several viral proteins have been found to block immune recognition of viral proteins as antigens during lytic replication, such as BGLF5, BZLF2, BILF1 and BNLF2a (Ressing, van Leeuwen et al. 2005; Rowe, Glaunsinger et al. 2007; Zuo, Thomas et al. 2008; Croft, Shannon-Lowe et al. 2009; Zuo, Currin et al. 2009; Zuo, Quinn et al. 2011). The direct contribution of BNFL2a in immune evasion was evidenced using an EBV BAC which initially disrupted the BNLF2a gene of the B95.8 strain by insertional mutagenesis (Croft, Shannon-Lowe et al. 2009). BNLF2 inhibits transporter associated with antigen (TAP). It encodes a 60 amino acid protein which prevents both peptide- and ATP-binding to TAP complex (Hislop, Ressing et al. 2007). Consequently, when co-expressed with target-antigens, cells expressing BNLF2a show decreased levels of surface human leukocyte antigen (HLA)-class I and are resistant to CD8+ cytotoxic T cell killing (Hislop, Ressing et al. 2007). Croft et al (Croft, Shannon-Lowe et al. 2009) created a targeting plasmid with BNLF2a gene which was replaced by tetracycline resistant cassette. This plasmid was then flanked by FLP recombinase target (FRT) sites. This vector was homologously recombined with the EBV BAC, and designated as ΔDBNLF2a, which had the tetracycline gene removed by FLP recombinase. ΔDBNLF2a BACs were then stably transduced into 293 cells, virus replication induced by transfection of a plasmid encoding the EBV lytic switch protein BZLF1 (Feederle, Kost et al. 2000). Compared to wild-type EBV BAC, this recombinant virus induces a strong MHC I T cell response against viral lytic genes than the wild type viruses (Feederle, Bartlett et al. 2010). Overall, these results indicate that BNLF2 prevents the immediate early and early proteins from being efficiently processed and presented to CD8 + T cells during lytic cycle replication. Contrary to BNLF2a in early evasion mechanism in the lytic cycle of EBV, other mechanism seems to operate later during immune evasion (Croft, Shannon-Lowe et al. 2009). Such stage-specific expression of immune evasion genes are a feature of several herpesviruses, such as CMV (Croft, Shannon-Lowe et al. 2009). Taken together, BNLF2a acts in concert with other immune-evasion genes encoded by EBV T-cell surveillance (Croft, Shannon-Lowe et al. 2009).

3. Non-viral gene functional studies

3.1 Translational research by the analysis of entire cancer genomes using BAC arrays

The development of high-resolution microarray-based comparative genomic hybridization (aCGH) using cDNA of BACs makes it possible for translational research to analyze the entire cancer genome in a single experiment. Well-designed aCGH studies will increase our understanding of the genetic basis of cancer, help to identify novel predictive and prognostic biomarkers for cancer, and molecular therapeutic targets in cancer. Compared to oligonucleotide arrays, BAC arrays have some specific features. BACs have been widely used in aCGH studies (Pinkel and Albertson 2005; Lockwood, Chari et al. 2006; Ylstra, van den Ijssel et al. 2006). The vast majority of array CGH data available today has been generated using BAC CGH arrays. BACs probes vary in length from 150 to 200 kb (Pinkel and Albertson 2005). The probe of genome-wide BAC arrays range from 2,400 to 32, 000 unique elements in tiling path array, where each BAC overlaps with its contiguous BACs. The resolution (the distance between each DNA target represented on the array) of each BAC array is defined by the number of unique probes it contains (Tan, Lambros et al. 2007). BAC tiling path arrays provide a resolution of up to 50 kb (Tan, Lambros et al. 2007). The development of a whole-genome BAC tiling path approach has improved resolution of CGH by using overlapping clones (Ishkanian, Malloff et al. 2004; Lockwood, Chari et al. 2006). These platforms provide sufficiently strong signals to detect single-copy change, and are able to accurately define the boundaries of genomic aberrations, which can possibly be utilized in archival formalin-fixed, paraffin-embedded (FFPE) tissue (Johnson, Hamoudi et al. 2006; Little, Vuononvirta et al. 2006).

High amounts of high-quality BAC DNA are needed to obtain good array performance (Ylstra, van den Ijssel et al. 2006). BACs DNA yield is generally low when isolated from *E. coli* (Pinkel and Albertson 2005). Because of the low yields of DNA from isolated BAC clones, DNA amplification is required to generate sufficient quantities of adequately pure BAC DNA for the assay. Therefore a tiling path array is costly and highly labor intensive. In addition, as BAC probes are representative of the human genome, they also contain repetitive sequences, which can result in nonspecific hybridization (Tan, Lambros et al. 2007).

3.2 Measurement of neuroblastoma DNA copy number aberrations (CNAs)

BAC technique has increasing been applied in detecting structural changes in chromosomes, such as copy number aberrations (CANs) and rearrangement. Mosse et al (Mosse, Greshock et al. 2005) generated 4135 BAC clones spanning the human genome at about 1.0 Mb resolution as targets for array-based comparative genomic hybridization (aCGH) experiments (Greshock, Naylor et al. 2004). They measured the relevance of neuroblastoma DNA copy number changes (CNAs) in forty-two human neuroblastoma cell lines. They found that all cell lines exhibited CNAs ranging from 2% to 41% of the genome. Chromosome 17 showed the highest frequency of CANs. The most frequent region of gain with high-level amplification localized to 2p24.22-2p24.3 detected in 81% of cell lines (Mosse, Greshock et al. 2005). Potential oncogenes such as MYCN, NAG and DDX1 were located in these regions. The less frequent region of gain localized to 17q23.2, 17q23.3-17q24.1, 17q24.1-17q24.2, 17q25.2-17q25.3 was detected in about 70% of the cell lines. Potential oncogene BIRC5 was localized in these regions. Although gain of 17q material was common, this low level gain of chromosomal material was rather complicated (Mosse, Greshock et al. 2005). The most frequent hemizygous deletion localized to a 4.0 Mb region at 1p36.23-36.32, was detected in 60% of the cell lines. Potential tumor suppressor genes TP73, CHD5, RPL22 and HKR3 were also localized. A 10.4 Mb region at 11q23.3-11q25 was detected in 36% of the cell lines, and the potential tumor suppressor gene CHEK1 was found there as well (Mosse, Greshock et al. 2005). Overall, the array CGH could be reliable in examining DNA copy number aberrations including single copy gain or loss. Compared to the data with standard techniques, data from array CGH correlates well with known aberrations detected by standard techniques. Therefore, array CGH can be applied to identify novel regions of genomic imbalance.

3.3 Fluorescence in situ hybridization (FISH) analysis of pathological archives with BACs

It is now known that there are extensive somatic changes, including multiple point mutations (Wood, Parsons et al. 2007; Velculescu 2008), copy number alterations (Weir, Woo et al. 2007; Kubo, Kuroda et al. 2009), and further complex rearrangements (Campbell, Stephens et al. 2008) tumors. But when and where these genetic changes occur during human cancer development remains unclear. Human archival tissue blocks contain specimens of human tumors in various stages of development, which are precious in the post-human-genome-sequencing era. Based on their findings and other's work, Sugimura et al (Sugimura, Mori et al. 2010) stated that the intensive application BAC clones as probes for FISH that have exact 'addresses' in the whole genome will become a useful diagnostic tools for pathologists. Thousands of BAC clones are commercially available, and any of them can be used as FISH probes. Sugimura et al tested 100 BAC probes containing different kinase loci in a gastric, colorectal, and lung cancer detection sets (20 cases for each organ) by using tissue microarray (TMA)-FISH technology (Sugimura, Mori et al. 2010). Sugimura et al found that unexpected kinase loci were amplified in a significant proportion of human common solid tumors (Sugimura, Mori et al. 2010). Combinatory chemistry has generated many drugs by targeting kinase genes or their products. Thus, amplification of specific regions on certain kinase genes are amenable to pharmacological intervention which could result in the target specific therapy. Therefore it is reasonable to believe that the FISH-BACs diagnostic system combined with particular kinase probes may provide the practical basis of individual cancer therapy.

3.4 Interferon-γ locus regulation with BACs

To investigate the regulatory properties of conserved non-coding sequence (CNS) element of interferon-γ (Ifng) gene *in vivo*, Hatton et al (Hatton, Harrington et al. 2006) developed a BAC-based transgenic reporter system to express Ifng gene expression. They introduced a Thy1.1 reporter into exon 1 of Ifng and placed this reporter into a BAC containing approximately 60 kb upstream of exon 1 of ifng and approximately 100 kb downstream of exon 4 of ifng sequences. The CNS-22 region of BAC was then flanked with loxP sites (Hatton, Harrington et al. 2006; Wilson and Schoenborn 2006). Because activation of the large BAC transgenic allele unlikely perturbs endogenous alleles (Valjent, Bertran-Gonzalez et al. 2009), potential confounding effects of altered IFNg production were possibly eliminated (Hatton, Harrington et al. 2006). Hatton et al chose Thy1.1 (CD90.1) as a reporter because of its low immunogenicity and easy detection in the context of CD90.2 allotype of the C57BL/6 background. The recombined BACs (Ifng-Thy1.1 BAC) containing the Thy1.1 reporter and floxed CNS-22 were microinjected into fertilized C57BL/6 oocytes (Hatton, Harrington et al. 2006; Wilson and Schoenborn 2006). As a result, the Ifng-Thy1.1 BAC-in transgene completely mirrored endogenous Ifng gene expression; and conditional deletion of the CNS-22 element from the single copy transgene by Cre recombinase resulted in almost complete loss of Thy1.1 expression in Th1 cells, CD8+ T cells, and NK cells irrespective of activation through the T cell receptor (TCR)-dependent or TCR-independent pathways. Thus, CNS-22 is considered to be critically involved in Ifng gene expression, irrespective of adaptive or innate immune cell lineage (Hatton, Harrington et al. 2006; Wilson and Schoenborn 2006). CNS-22 functions as an enhancer both *in vitro* and *in vivo*, which will shed new light on ifng regulation and open up avenues for future investigation.

3.5 Studies of Kras-mediated pancreatic tumorigenesis with BACs

Activation of Kras gene by mutation plays a critical role in human pancreatic cancer (Almoguera, Shibata et al. 1988; Shibata, Almoguera et al. 1990). Although the known capability of oncogenic Kras to function as a key initiator of pancreatic malignancy, the mechanism(s) of Kras-caused initiating events are still unclear. This is an important reason why prognosis for patients with malignant pancreatic tumors have not entirely improved over the past twenty years (Jemal, Siegel et al. 2006). Moore et al have discovered that the zebrafish develop pancreatic cancer after exposure to chemical mutagens (Moore, Rush et al. 2006). The studies have also shown that mammalian and zebrafish pancreas are significantly similar in anatomy and histology (Wallace and Pack 2003; Chen, Li et al. 2007). Therefore, the Zebrafish has emerged as an experimental model for study of human pancreatic cancer biology (Davison, Woo Park et al. 2008; Park, Davison et al. 2008). Another benefits of working with zebrafish model is their translucency, which greatly improves the visualization of fluorescent trangenes in both embryos and adult zebrafish (Davison, Woo Park et al. 2008). Park et al (Park, Davison et al. 2008) discovered that oncogenic Kras causes pancreatic cell expansion and malignant transformation in the zebrafish exocrine pancreas by utilizing eGFP-Kras BAC transgenes (160kb) under the regulation of Ptf1a regulatory elements. Ptf1a induces differentiation, growth and proliferation of pancreatic progenitor cells (Park, Davison et al. 2008). Briefly, they expressed either extended green fluorescent protein (eGFP) alone or eGFP fused to oncogenic Kras in developing zebrafish pancreas and continuously detected the expression of fluorescent transgenes transcutaneously during all stages of development including the adult zebrafish. They first generated polymerase chain reaction (PCR) products encoding the eGFP and eGFP-Kras transgenes flanked by sequences homologous to the CH211-142 BAC that spans the Pfta1 gene locus. Homologous recombination leads to accurate replacement of the Ptf1a coding sequences with the eGFP and eGFP-Kras transgene (Davison, Woo Park et al. 2008). Their results demonstrate that oncogenic Kras-expressed pancreatic progenitor cells fail to undergo characteristic exocrine differentiation although their initial specification and migration are observed to be normal (Davison, Woo Park et al. 2008; Park, Davison et al. 2008). Blocks of differentiation leads to abnormal accumulation of the undifferentiated progenitor cells, correlates with the formation of invasive pancreatic cancer. Besides similarity in anatomy and histology, Zebrafish pancreatic tumors share several activated signaling pathways with the human pancreatic tumors, including activation of ERK and AKTby phosphorylation, as well as abnormal Hedgehog pathway activation which was justified by the up-regulation of ptc1 mRNA and gli1 mRNA (Park, Davison et al. 2008). These findings provide a unique view of the tumor-initiating effects of oncogenic Kras in a living vertebrate organism, but more important it suggest that BACs transgene targeting other oncogenes or tumor suppressor genes in zebrafish pancreatic cancer may improve our understanding of the human disease.

3.6 Studies on striatal signaling pathways in central nervous system (CNS) with BACs

To understand the role of molecular signaling pathways involved in behavioral responses, it is necessary to delineate the molecular events that take place in neurons. This task has been hampered by the complexity of neuronal system. There are hundreds of distinct neuronal populations and these populations are very difficult to distinguish (Valjent, Bertran-Gonzalez et al. 2009). The development of BAC transgenic mice expressing various reporters, epitope tagged-proteins or Cre recombinase driven by specific promoters, greatly facilitates the research in this field. Generally speaking, transgene expression is influenced by copy numbers and site of insertions (positional effects). Large BAC transgenes (BACs contain large fragments 150-200kb of mouse genome) have usually a low copy number, are less likely influenced by positional effects, and are able to recapitulate the regulation of endogenous genes much better than shorter transgenes (Yang, Model et al. 1997). Over the past few years, the use of BAC EGFP transgenic mice have generated significant development in the analysis of striatal physiology and physiopathology (Valjent, Bertran-Gonzalez et al. 2009). The drd1a-EGFP (EGFP reporter is driven by dopamine D1 receptor-D1R promoter), drd2-EGFP (EGFP reporter is driven by dopamine D2 receptor-D2R promoter) and chrm4-EGFP (EGFP reporter is driven by cholinergic receptor, muscarinic 4-CHRM4 receptor promoter) BAC transgenic mice have been extensively utilized to investigate the physiological features of striatonigral and striatopallidal medium spiny projection neurons (MSNs) (Lobo, Karsten et al. 2006; Kreitzer and Malenka 2007; Cepeda, Andre et al. 2008; Gertler, Chan et al. 2008). Among these major findings are as follows: 1)D1R-expressing MSNs are less excitable than D2R-MSNs (Lobo, Karsten et al. 2006; Kreitzer and Malenka 2007; Cepeda, Andre et al. 2008; Gertler, Chan et al. 2008) due to different morphology (Gertler, Chan et al. 2008), and some presynaptic factors (Kreitzer and Malenka 2007; Cepeda, Andre et al. 2008). Corticostriatal synapses are activated by repetitive stimulation; in contrast, thalamostriatal synapses are inhibited by repetitive stimulation (Ding, Peterson et al. 2008); 2) D1R-expressing MSNs collaterals are functionally connected primarily with other D1R-MSNs, whereas D2R-expressing neurons collaterals are connected with both D2R- and D1R-MSNs (Taverna, Ilijic et al. 2008). D2R-MSNs synapse with GABAA receptors are stronger (Taverna, Ilijic et al. 2008) and generate greater GABAA receptor-mediated tonic currents (Ade, Janssen et al. 2008) than D1R-MSNs (Janssen, Ade et al. 2009); 3)The single back-propagating action potentials invade more distal dendritic regions in D2R-than in D1R-MSNs, due to a difference in voltage-dependent Na+channels and Kv4 K+ channels (Day, Wokosin et al. 2008); 4) In the dopamine-depleted striatum, the corticostriatal connections are decreased in D2R neurons (Day, Wang et al. 2006), whereas dendritic excitability is increased in this region (Day, Wokosin et al. 2008; Taverna, Ilijic et al. 2008).

4. Conclusions and overall perspectives

Studies on BACs have demonstrated their importance in many research fields, from microbiology, virology, to human genetics, neuroscience, and proteomics (Narayanan 2008; Adamson, Jackson et al. 2011). The power to clone and handle large sized intact genome with high fidelity by BAC has enabled scientists to design and perform both mechanistic and functional studies in an ever expanding field.

5. Acknowledgements

This review was supported by the National Institutes of Health grant numbers NIH-FIC (1T90-HG004151-01) for postdoctoral training in Genomics and Hemoglobinopathies, NIH/FIC/NINDS R21 and NIH-RCMI (RR033062).

6. Glossary

aCGH: array-based comparative genomic hybridization
BACs: bacterial artificial chromosome
BHV-1: bovine herpesvirus Type 1
CGH: comparative genomic hybridization
CHRM4: cholinergic receptor, muscarinic 4- receptor
CMV: cytomegalovirus
CNAs: copy number changes
CNS: central nervous system

D1R: dopamine D1 receptor
D2R: dopamine D2 receptor
ECs: endothelial cells
eGFP: extended green fluorescent protein
EHV-1: Equine herpesvirus Type 1
FFPE: archival formalin-fixed, paraffin-embedded
FHV-1: Feline herpesvirus
FISH: fluorescence in situ hybridization
F-plasmid: a fertility plasmid
GPCMV: Guinea pig cytomegalovirus
HCMV or HHV-5: human cytomegalovirus
HFF: human foreskin fibroblasts
HLA: human leukocyte antigen
HSV-1: Herpes simples virus type 1
HSV-2: Herpes simples virus type 2
HUVECs: human umbilical vein endothelial cells
HVS: herpesvirus saimiri
HVT: Turkery herpesvirus
HPV: herpesviruses
Ifng: interferon-γ
KHV: Koi herpesvirus
KSHV or HHV-8: Kaposi's sarcoma-associated herpesvirus,
LCV: lymphocryptovirus
MDV: Marek's disease virus
mECK36 cells: mouse bone marrow endothelial cells generated by transfection of with KSHVBac36
MSNs: medium spiny projection neurons
VEGF: vascular endothelial growth factor
vGPCR: viral G protein-coupled receptor
VZV or HHV-3: varicella-zoster virus,
KSHVBac36: KSHV bacterial artificial chromosome,
mCMV: murine gammaherpesvirus
MHV-68: Murine cytomegalovirus 68
ORFs: open reading frames
PCR: polymerase chain reaction
PrV: pseudorabies virus
rhCMV: Rhesus cytomegalovirus
RRV: rhesus rhadinovirus
TMA: tissue microarray

7. References

Adamson, A. D., D. Jackson, et al. (2011). "Novel approaches to in vitro transgenesis." J Endocrinol 208(3): 193-206.
Ade, K. K., M. J. Janssen, et al. (2008). "Differential tonic GABA conductances in striatal medium spiny neurons." J Neurosci 28(5): 1185-1197.

Almoguera, C., D. Shibata, et al. (1988). "Most human carcinomas of the exocrine pancreas contain mutant c-K-ras genes." Cell 53(4): 549-554.

Altmann, M. and W. Hammerschmidt (2005). "Epstein-Barr virus provides a new paradigm: a requirement for the immediate inhibition of apoptosis." PLoS Biol 3(12): e404.

Altmann, M., D. Pich, et al. (2006). "Transcriptional activation by EBV nuclear antigen 1 is essential for the expression of EBV's transforming genes." Proc Natl Acad Sci U S A 103(38): 14188-14193.

Bissinger, A. L., C. Sinzger, et al. (2002). "Human cytomegalovirus as a direct pathogen: correlation of multiorgan involvement and cell distribution with clinical and pathological findings in a case of congenital inclusion disease." J Med Virol 67(2): 200-206.

Borenstein, R. and N. Frenkel (2009). "Cloning human herpes virus 6A genome into bacterial artificial chromosomes and study of DNA replication intermediates." Proc Natl Acad Sci U S A 106(45): 19138-19143.

Borst, E. M., G. Hahn, et al. (1999). "Cloning of the human cytomegalovirus (HCMV) genome as an infectious bacterial artificial chromosome in Escherichia coli: a new approach for construction of HCMV mutants." J Virol 73(10): 8320-8329.

Brune, W., C. Menard, et al. (2001). "A ribonucleotide reductase homolog of cytomegalovirus and endothelial cell tropism." Science 291(5502): 303-305.

Campbell, P. J., P. J. Stephens, et al. (2008). "Identification of somatically acquired rearrangements in cancer using genome-wide massively parallel paired-end sequencing." Nat Genet 40(6): 722-729.

Cepeda, C., V. M. Andre, et al. (2008). "Differential electrophysiological properties of dopamine D1 and D2 receptor-containing striatal medium-sized spiny neurons." Eur J Neurosci 27(3): 671-682.

Chen, A., M. Divisconte, et al. (2005). "Epstein-Barr virus with the latent infection nuclear antigen 3B completely deleted is still competent for B-cell growth transformation in vitro." J Virol 79(7): 4506-4509.

Chen, S., C. Li, et al. (2007). "Anatomical and histological observation on the pancreas in adult zebrafish." Pancreas 34(1): 120-125.

Copeland, N. G., N. A. Jenkins, et al. (2001). "Recombineering: a powerful new tool for mouse functional genomics." Nat Rev Genet 2(10): 769-779.

Costa, J. L., G. Meijer, et al. (2008). "Array comparative genomic hybridization copy number profiling: a new tool for translational research in solid malignancies." Semin Radiat Oncol 18(2): 98-104.

Cowell, J. K. (2004). "High throughput determination of gains and losses of genetic material using high resolution BAC arrays and comparative genomic hybridization." Comb Chem High Throughput Screen 7(6): 587-596.

Cowell, J. K. and N. J. Nowak (2003). "High-resolution analysis of genetic events in cancer cells using bacterial artificial chromosome arrays and comparative genome hybridization." Adv Cancer Res 90: 91-125.

Croft, N. P., C. Shannon-Lowe, et al. (2009). "Stage-specific inhibition of MHC class I presentation by the Epstein-Barr virus BNLF2a protein during virus lytic cycle." PLoS Pathog 5(6): e1000490.

Dankner, W. M., J. A. McCutchan, et al. (1990). "Localization of human cytomegalovirus in peripheral blood leukocytes by in situ hybridization." J Infect Dis 161(1): 31-36.

Davison, J. M., S. Woo Park, et al. (2008). "Characterization of Kras-mediated pancreatic tumorigenesis in zebrafish." Methods Enzymol 438: 391-417.

Day, M., Z. Wang, et al. (2006). "Selective elimination of glutamatergic synapses on striatopallidal neurons in Parkinson disease models." Nat Neurosci 9(2): 251-259.

Day, M., D. Wokosin, et al. (2008). "Differential excitability and modulation of striatal medium spiny neuron dendrites." J Neurosci 28(45): 11603-11614.

Delecluse, H. J., T. Hilsendegen, et al. (1998). "Propagation and recovery of intact, infectious Epstein-Barr virus from prokaryotic to human cells." Proc Natl Acad Sci U S A 95(14): 8245-8250.

Delecluse, H. J., M. Kost, et al. (2001). "Spontaneous activation of the lytic cycle in cells infected with a recombinant Kaposi's sarcoma-associated virus." J Virol 75(6): 2921-2928.

Ding, J., J. D. Peterson, et al. (2008). "Corticostriatal and thalamostriatal synapses have distinctive properties." J Neurosci 28(25): 6483-6492.

Dulal, K., Z. Zhang, et al. (2009). "Development of a gene capture method to rescue a large deletion mutant of human cytomegalovirus." J Virol Methods 157(2): 180-187.

Feederle, R., E. J. Bartlett, et al. (2010). "Epstein-Barr virus genetics: talking about the BAC generation." Herpesviridae 1(1): 6.

Feederle, R., M. Kost, et al. (2000). "The Epstein-Barr virus lytic program is controlled by the co-operative functions of two transactivators." Embo J 19(12): 3080-3089.

Gertler, T. S., C. S. Chan, et al. (2008). "Dichotomous anatomical properties of adult striatal medium spiny neurons." J Neurosci 28(43): 10814-10824.

Gnann, J. W., Jr., J. Ahlmen, et al. (1988). "Inflammatory cells in transplanted kidneys are infected by human cytomegalovirus." Am J Pathol 132(2): 239-248.

Greshock, J., T. L. Naylor, et al. (2004). "1-Mb resolution array-based comparative genomic hybridization using a BAC clone set optimized for cancer gene analysis." Genome Res 14(1): 179-187.

Hahn, G., H. Khan, et al. (2002). "The human cytomegalovirus ribonucleotide reductase homolog UL45 is dispensable for growth in endothelial cells, as determined by a BAC-cloned clinical isolate of human cytomegalovirus with preserved wild-type characteristics." J Virol 76(18): 9551-9555.

Hahn, G., M. G. Revello, et al. (2004). "Human cytomegalovirus UL131-128 genes are indispensable for virus growth in endothelial cells and virus transfer to leukocytes." J Virol 78(18): 10023-10033.

Halder, S., M. Murakami, et al. (2009). "Early events associated with infection of Epstein-Barr virus infection of primary B-cells." PLoS One 4(9): e7214.

Hatton, R. D., L. E. Harrington, et al. (2006). "A distal conserved sequence element controls Ifng gene expression by T cells and NK cells." Immunity 25(5): 717-729.

Hislop, A. D., M. E. Ressing, et al. (2007). "A CD8+ T cell immune evasion protein specific to Epstein-Barr virus and its close relatives in Old World primates." J Exp Med 204(8): 1863-1873.

Horsburgh, B. C., M. M. Hubinette, et al. (1999). "Allele replacement: an application that permits rapid manipulation of herpes simplex virus type 1 genomes." Gene Ther 6(5): 922-930.

Howell, C. L., M. J. Miller, et al. (1979). "Comparison of rates of virus isolation from leukocyte populations separated from blood by conventional and Ficoll-Paque/Macrodex methods." J Clin Microbiol 10(4): 533-537.

Hutt-Fletcher, L. M. (2007). "Epstein-Barr virus entry." J Virol 81(15): 7825-7832.

Imelfort, M., J. Batley, et al. (2009). "Genome sequencing approaches and successes." Methods Mol Biol 513: 345-358.

Ishkanian, A. S., C. A. Malloff, et al. (2004). "A tiling resolution DNA microarray with complete coverage of the human genome." Nat Genet 36(3): 299-303.

Izumi, K. M. (2001). "Identification of EBV transforming genes by recombinant EBV technology." Semin Cancer Biol 11(6): 407-414.

Jackson, R. J. and N. Standart (2007). "How do microRNAs regulate gene expression?" Sci STKE 2007(367): re1.

Janes, D. E., N. Valenzuela, et al. (2011). "Sex chromosome evolution in amniotes: applications for bacterial artificial chromosome libraries." J Biomed Biotechnol 2011(12): 132975.

Janssen, M. J., K. K. Ade, et al. (2009). "Dopamine modulation of GABA tonic conductance in striatal output neurons." J Neurosci 29(16): 5116-5126.

Jarvis, M. A. and J. A. Nelson (2007). "Human cytomegalovirus tropism for endothelial cells: not all endothelial cells are created equal." J Virol 81(5): 2095-2101.

Jemal, A., R. Siegel, et al. (2006). "Cancer statistics, 2006." CA Cancer J Clin 56(2): 106-130.

Johnson, N. A., R. A. Hamoudi, et al. (2006). "Application of array CGH on archival formalin-fixed paraffin-embedded tissues including small numbers of microdissected cells." Lab Invest 86(9): 968-978.

Jones, B. K., B. R. Monks, et al. (1995). "The human growth hormone gene is regulated by a multicomponent locus control region." Mol Cell Biol 15(12): 7010-7021.

Kahl, M., D. Siegel-Axel, et al. (2000). "Efficient lytic infection of human arterial endothelial cells by human cytomegalovirus strains." J Virol 74(16): 7628-7635.

Kanda, T., M. Yajima, et al. (2004). "Production of high-titer Epstein-Barr virus recombinants derived from Akata cells by using a bacterial artificial chromosome system." J Virol 78(13): 7004-7015.

Kempkes, B., D. Pich, et al. (1995). "Immortalization of human B lymphocytes by a plasmid containing 71 kilobase pairs of Epstein-Barr virus DNA." J Virol 69(1): 231-238.

Knipe, D. M., W. Batterson, et al. (1981). "Molecular genetics of herpes simplex virus. VI. Characterization of a temperature-sensitive mutant defective in the expression of all early viral gene products." J Virol 38(2): 539-547.

Kreitzer, A. C. and R. C. Malenka (2007). "Endocannabinoid-mediated rescue of striatal LTD and motor deficits in Parkinson's disease models." Nature 445(7128): 643-647.

Kubo, T., Y. Kuroda, et al. (2009). "Resequencing and copy number analysis of the human tyrosine kinase gene family in poorly differentiated gastric cancer." Carcinogenesis 30(11): 1857-1864.

Lander, E. S., L. M. Linton, et al. (2001). "Initial sequencing and analysis of the human genome." Nature 409(6822): 860-921.

Little, S. E., R. Vuononvirta, et al. (2006). "Array CGH using whole genome amplification of fresh-frozen and formalin-fixed, paraffin-embedded tumor DNA." Genomics 87(2): 298-306.

Liu, M., S. Guo. (2010). "p38 and STAT3 activation by vGPCR in KSHV-infected cells." Virus Adaptation and Treatment(2): 103–113.

Lobo, M. K., S. L. Karsten, et al. (2006). "FACS-array profiling of striatal projection neuron subtypes in juvenile and adult mouse brains." Nat Neurosci 9(3): 443-452.

Lockwood, W. W., R. Chari, et al. (2006). "Recent advances in array comparative genomic hybridization technologies and their applications in human genetics." Eur J Hum Genet 14(2): 139-148.

Lu, X. H. (2009). "BAC to degeneration bacterial artificial chromosome (BAC)-mediated transgenesis for modeling basal ganglia neurodegenerative disorders." Int Rev Neurobiol 89: 37-56.

MacCormac, L. P. and J. E. Grundy (1999). "Two clinical isolates and the Toledo strain of cytomegalovirus contain endothelial cell tropic variants that are not present in the AD169, Towne, or Davis strains." J Med Virol 57(3): 298-307.

Magin-Lachmann, C., G. Kotzamanis, et al. (2004). "In vitro and in vivo delivery of intact BAC DNA -- comparison of different methods." J Gene Med 6(2): 195-209.

Marchini, A., H. Liu, et al. (2001). "Human cytomegalovirus with IE-2 (UL122) deleted fails to express early lytic genes." J Virol 75(4): 1870-1878.

Meseda, C. A., F. Schmeisser, et al. (2004). "DNA immunization with a herpes simplex virus 2 bacterial artificial chromosome." Virology 318(1): 420-428.

Moore, J. L., L. M. Rush, et al. (2006). "Zebrafish genomic instability mutants and cancer susceptibility." Genetics 174(2): 585-600.

Mosse, Y. P., J. Greshock, et al. (2005). "Measurement and relevance of neuroblastoma DNA copy number changes in the post-genome era." Cancer Lett 228(1-2): 83-90.

Murphy, E., D. Yu, et al. (2003). "Coding potential of laboratory and clinical strains of human cytomegalovirus." Proc Natl Acad Sci U S A 100(25): 14976-14981.

Mutlu, A. D., L. E. Cavallin, et al. (2007). "In vivo-restricted and reversible malignancy induced by human herpesvirus-8 KSHV: a cell and animal model of virally induced Kaposi's sarcoma." Cancer Cell 11(3): 245-258.

Myerson, D., R. C. Hackman, et al. (1984). "Widespread presence of histologically occult cytomegalovirus." Hum Pathol 15(5): 430-439.

Nagaike, K., Y. Mori, et al. (2004). "Cloning of the varicella-zoster virus genome as an infectious bacterial artificial chromosome in Escherichia coli." Vaccine 22(29-30): 4069-4074.

Narayanan, K. (2008). "Intact recombineering of highly repetitive DNA requires reduced induction of recombination enzymes and improved host viability." Anal Biochem 375(2): 394-396.

Park, S. W., J. M. Davison, et al. (2008). "Oncogenic KRAS induces progenitor cell expansion and malignant transformation in zebrafish exocrine pancreas." Gastroenterology 134(7): 2080-2090.

Pinkel, D. and D. G. Albertson (2005). "Array comparative genomic hybridization and its applications in cancer." Nat Genet 37 Suppl(7): S11-17.

Read, R. W., J. A. Zhang, et al. (1999). "Evaluation of the role of human retinal vascular endothelial cells in the pathogenesis of CMV retinitis." Ocul Immunol Inflamm 7(3-4): 139-146.

Ressing, M. E., D. van Leeuwen, et al. (2005). "Epstein-Barr virus gp42 is posttranslationally modified to produce soluble gp42 that mediates HLA class II immune evasion." J Virol 79(2): 841-852.

Rowe, M., B. Glaunsinger, et al. (2007). "Host shutoff during productive Epstein-Barr virus infection is mediated by BGLF5 and may contribute to immune evasion." Proc Natl Acad Sci U S A 104(9): 3366-3371.

Saeki, Y., T. Ichikawa, et al. (1998). "Herpes simplex virus type 1 DNA amplified as bacterial artificial chromosome in Escherichia coli: rescue of replication-competent virus progeny and packaging of amplicon vectors." Hum Gene Ther 9(18): 2787-2794.

Schalkwyk, L. C., F. Francis, et al. (1995). "Techniques in mammalian genome mapping." Curr Opin Biotechnol 6(1): 37-43.

Scrivano, L., C. Sinzger, et al. (2011). "HCMV spread and cell tropism are determined by distinct virus populations." PLoS Pathog 7(1): e1001256.

Semprini, S., T. J. Troup, et al. (2007). "Cryptic loxP sites in mammalian genomes: genome-wide distribution and relevance for the efficiency of BAC/PAC recombineering techniques." Nucleic Acids Res 35(5): 1402-1410.

Shibata, D., C. Almoguera, et al. (1990). "Detection of c-K-ras mutations in fine needle aspirates from human pancreatic adenocarcinomas." Cancer Res 50(4): 1279-1283.

Shizuya, H., B. Birren, et al. (1992). "Cloning and stable maintenance of 300-kilobase-pair fragments of human DNA in Escherichia coli using an F-factor-based vector." Proc Natl Acad Sci U S A 89(18): 8794-8797.

Sinzger, C., A. Grefte, et al. (1995). "Fibroblasts, epithelial cells, endothelial cells and smooth muscle cells are major targets of human cytomegalovirus infection in lung and gastrointestinal tissues." J Gen Virol 76 (Pt 4)(Pt 4): 741-750.

Sinzger, C., G. Hahn, et al. (2008). "Cloning and sequencing of a highly productive, endotheliotropic virus strain derived from human cytomegalovirus TB40/E." J Gen Virol 89(Pt 2): 359-368.

Sinzger, C., K. Schmidt, et al. (1999). "Modification of human cytomegalovirus tropism through propagation in vitro is associated with changes in the viral genome." J Gen Virol 80 (Pt 11)(Pt 11): 2867-2877.

Sparwasser, T. and G. Eberl (2007). "BAC to immunology--bacterial artificial chromosome-mediated transgenesis for targeting of immune cells." Immunology 121(3): 308-313.

Stavropoulos, T. A. and C. A. Strathdee (1998). "An enhanced packaging system for helper-dependent herpes simplex virus vectors." J Virol 72(9): 7137-7143.

Sugimura, H., H. Mori, et al. (2010). "Fluorescence in situ hybridization analysis with a tissue microarray: 'FISH and chips' analysis of pathology archives." Pathol Int 60(8): 543-550.

Tan, D. S., M. B. Lambros, et al. (2007). "Getting it right: designing microarray (and not 'microawry') comparative genomic hybridization studies for cancer research." Lab Invest 87(8): 737-754.

Taverna, S., E. Ilijic, et al. (2008). "Recurrent collateral connections of striatal medium spiny neurons are disrupted in models of Parkinson's disease." J Neurosci 28(21): 5504-5512.

Townes, T. M., J. B. Lingrel, et al. (1985). "Erythroid-specific expression of human beta-globin genes in transgenic mice." Embo J 4(7): 1715-1723.

Tunster, S. J., M. Van De Pette, et al. (2011). "BACs as tools for the study of genomic imprinting." J Biomed Biotechnol 2011(13): 283013.

Valjent, E., J. Bertran-Gonzalez, et al. (2009). "Looking BAC at striatal signaling: cell-specific analysis in new transgenic mice." Trends Neurosci 32(10): 538-547.

Velculescu, V. E. (2008). "Defining the blueprint of the cancer genome." Carcinogenesis 29(6): 1087-1091.

Venter, J. C., M. D. Adams, et al. (2001). "The sequence of the human genome." Science 291(5507): 1304-1351.

Wallace, K. N. and M. Pack (2003). "Unique and conserved aspects of gut development in zebrafish." Dev Biol 255(1): 12-29.

Wang, D., Q. C. Yu, et al. (2007). "Human cytomegalovirus uses two distinct pathways to enter retinal pigmented epithelial cells." Proc Natl Acad Sci U S A 104(50): 20037-20042.

Warden, C., Q. Tang, et al. (2011). "Herpesvirus BACs: past, present, and future." J Biomed Biotechnol 2011(27): 124595.

Weir, B. A., M. S. Woo, et al. (2007). "Characterizing the cancer genome in lung adenocarcinoma." Nature 450(7171): 893-898.

Wiley, C. A. and J. A. Nelson (1988). "Role of human immunodeficiency virus and cytomegalovirus in AIDS encephalitis." Am J Pathol 133(1): 73-81.

Wilson, C. B. and J. Schoenborn (2006). "BACing up the interferon-gamma locus." Immunity 25(5): 691-693.

Wood, L. D., D. W. Parsons, et al. (2007). "The genomic landscapes of human breast and colorectal cancers." Science 318(5853): 1108-1113.

Wussow, F., H. Fickenscher, et al. (2009). "Red-mediated transposition and final release of the mini-F vector of a cloned infectious herpesvirus genome." PLoS One 4(12): e8178.

Yang, X. W., P. Model, et al. (1997). "Homologous recombination based modification in Escherichia coli and germline transmission in transgenic mice of a bacterial artificial chromosome." Nat Biotechnol 15(9): 859-865.

Ylstra, B., P. van den Ijssel, et al. (2006). "BAC to the future! or oligonucleotides: a perspective for micro array comparative genomic hybridization (array CGH)." Nucleic Acids Res 34(2): 445-450.

Zhang, Z., J. Rowe, et al. (2007). "Genetic analysis of varicella-zoster virus ORF0 to ORF4 by use of a novel luciferase bacterial artificial chromosome system." J Virol 81(17): 9024-9033.

Zhou, F. C., Y. J. Zhang, et al. (2002). "Efficient infection by a recombinant Kaposi's sarcoma-associated herpesvirus cloned in a bacterial artificial chromosome: application for genetic analysis." J Virol 76(12): 6185-6196.

Zhou, G. and B. Roizman (2005). "Characterization of a recombinant herpes simplex virus 1 designed to enter cells via the IL13Ralpha2 receptor of malignant glioma cells." J Virol 79(9): 5272-5277.

Zuo, J., A. Currin, et al. (2009). "The Epstein-Barr virus G-protein-coupled receptor contributes to immune evasion by targeting MHC class I molecules for degradation." PLoS Pathog 5(1): e1000255.

Zuo, J., L. L. Quinn, et al. (2011). "The Epstein-Barr virus-encoded BILF1 protein modulates immune recognition of endogenously processed antigen by targeting major histocompatibility complex class I molecules trafficking on both the exocytic and endocytic pathways." J Virol 85(4): 1604-1614.

Zuo, J., W. Thomas, et al. (2008). "The DNase of gammaherpesviruses impairs recognition by virus-specific CD8+ T cells through an additional host shutoff function." J Virol 82(5): 2385-2393.

7

Bacterial Artificial Chromosome-Based Experimental Strategies in the Field of Developmental Neuroscience

Youhei W. Terakawa[1,2], Yukiko U. Inoue[1], Junko Asami[1] and Takayoshi Inoue[1]

[1]*Department of Biochemistry and Cellular Biology, National Institute of Neuroscience National Center of Neurology and Psychiatry, Kodaira, Tokyo*

[2]*Department of Electrical Engineering and Bioscience, Graduate School of Advanced Science and Engineering, Waseda University, Shinjuku-ku, Tokyo*

Japan

1. Introduction

Bacterial artificial chromosomes (BACs) constitute minimal components of various whole genome-sequencing projects, including our own. The recent innovation of efficient recombinogenic bacterial strains allows systematic BAC modifications (i.e. recombineering; Reviewed in Copeland et al., 2001), setting BACs as an ideal experimental basis for functional genomics with their broad coverage of transcriptional regulatory elements. For instance, in the field of neuroscience, the GENSAT project extensively modified mouse BAC clones, covering gene transcriptional units preferentially expressed in the nervous system, and generated hundreds of BAC transgenic (Tg) mouse lines from those modified BACs (Gong et al., 2003). These BAC-Tg lines successfully recapitulated complex gene expression profiles in the nervous system (Gong et al., 2003), providing a rigid analytical platform so as to be able to answer the fundamental question of how tens of millions of neurons and thousands of cell-types can become elaborate interconnected circuitries in the brain by using only twenty-thousand sets of gene transcriptional activities.

Classic cadherins are adhesion molecules at the cell-cell adherence junction and the neuron-neuron synapse peri-active zone (i.e. *puncta adherentia*) whose expression differentially delineates elaborated cytoarchitectures, such as layers and nuclei that constitute the basis for neural circuit formation in the vertebrate nervous system (Takeichi, 1995; Takeichi and Abe, 2005). Classic cadherins have 20 subclass members encoded by different genes, with each subclass harbouring distinct cell adhesiveness (Takeichi, 1995). In the *in vitro* aggregation assay system, it has been demonstrated that dissociated cells expressing the same sets of the classic cadherin subclass at a same level tend to make aggregates depending upon the calcium ion (Nose et al., 1988; Steinberg et al., 1994). Noticeably, in the developing nervous system, each classic cadherin subclass shows unique expression patterns, and such expression profiles are dynamically regulated during morphogenetic processes (Gumbiner, 2005; Redies, 2000). For instance, during chickens' early neural development, prospective neural tissue begins to express N-cadherin and, at the interface between the N-cadherin expressing cells and the surface ectodermal cells with E-cadherin expression, a Cadherin-6B positive, N-cadherin negative domain appears to segregate neural crest cells. Once they have emigrated from the neural plate/tube, neural crest cells finally begin to express Cadherin-7. This dynamical cadherin class switch is critical to the regulation of neurulation dynamics, since ectopically expressed N-cadherin perturbs the neural tube segregations from the ectoderm and/or the neural crest cell emigrations (Fujimori et al., 1990; Nakagawa & Takeichi, 1998). It is thus suggested that spatio-temporally regulated cadherin expression plays a pivotal role in animal morphogenetic processes.

While the physiological significance of differential cadherin expression profiles and/or cadherin class switches in neural development has been implicated, their gene regulatory mechanisms are largely unknown, due to the huge size and complex organisation of cadherin gene structures. Understanding the regulatory mechanisms for classic cadherin expression is crucial from a clinical point of view, as many cancer cells lose precise expression profiles of cadherins, resulting in the hyper-growth of cells and/or cell metastasis (Takeichi, 1995).

In our early studies, we have applied BAC-based technologies so as to screen gene regulatory patterns for a subclass of classic cadherins, cadherin-6 (*Cdh6*) whose gene structure is too large and complex for the identification of its promoter/enhancers by conventional methods. We succeeded in finding out that different genomic territories, located as far as 100-kbp upstream or downstream from the transcription start-site, is required for *Cdh6* expression at the defined time and place (Inoue et al., 2008a; Inoue et al., 2008b). Here, we extend the enhancer screening and reveal that a 6-kbp sized 3 prime intergenic region is critical in order to yield *Cdh6* expression along the somatosensory barrel, a distinct cytoarchitecture of the mouse cerebral cortex, at around postnatal day 7 (P7). Additionally, by taking advantage of the *Cdh6* enhancer/promoter activity identified, we establish a BAC Tg mouse line in which somatosensory barrels are stably illuminated by exogenous green fluorescent protein (GFP) expression, allowing us to suggest the roles of the retinoic acid (RA) related signalling pathway during cortical barrel field development and/or patterning. These results clearly demonstrate the strictly divisible *Cdh6* regulatory pattern along functional brain units, and the value of BAC-based experimental strategies in the field of developmental neuroscience.

2. Protocols

2.1 BAC modification via homologous recombination in bacterial cells

The recombinogenic bacterial strain *EL250* (Lee et al., 2001) was used in the present study. For the homologous recombination, ~1-kbp homology arms were amplified from the BAC clone of interest (=*RP23-78N21* in this study) by means of a polymerase chain reaction (PCR), and were cloned into a conventional TA cloning vector (pGEM-T-Easy; Promega). For homologous recombination, beta-galactosidase (*LacZ*) and a growth-associated protein-43 tagged enhanced green fluorescent protein (GAP43-EGFP) gene expression cassette were inserted into the BAC clone of interest; a *LacZ*/GAP-EGFP gene expression cassette and a gene cassette for clone selections (e.g. ampicillin resistant gene, kanamycin resistant gene, etc.) were cloned in between the homology arms and the fragment containing arms as well as reporter/selection gene cassettes which were purified by means of agarose gel electrophoresis after complete digestion with proper restriction enzymes. For homologous recombination mediated deletion of given territories from the BAC clone of interest, only the selection cassette was cloned in between the homology arms and the portion containing arms, and the selection cassette was isolated as described above. The selection cassette was put in between two *FRT* sites for its eventual excision by the inducible *Flpe* gene (see below).

The recombination system in *EL250* cells (Lee et al., 2001) was activated via heat shock, and their electroporation-competency (EP) was conferred on ice so as to maintain the recombinogenic activity. Briefly, *EL250* cells that harbour the BAC clone of interest were pre-cultured in 50 ml of LB without antibiotic at 32°C, until an optical density of 600 nm (OD_{600}) was reached at 0.5. By adding 2.5 ml of the overnight saturated culture to 50 ml of LB, it normally took ~90 min to obtain 0.5 OD_{600} in the shaking culture at 32°C. Then, 15 ml of the culture was heat-shocked at 42°C for 15 min in a water bath with the shaker's agitation, while another 15 ml was maintained in the shaking culture at 32°C as the control. Both of the cultures were subsequently placed on ice for 15min, centrifuged, and quickly washed three times by ice-cold water in order to draw the electro-competency for the heat-shocked cells. A few hundred nano-grams of the resultant purified fragment was then electroporated with a 1 mm gap cuvette by 1.8 kV (BioRad) into the EP-competent cells, pre-incubated for 2 hours at 32°C, and selected on an LB agar-plate containing 12.5 μg/ml chloramphenicol and 12.5 μg/ml Kanamycin/Ampicillin for ~40 hours at 32°C. Starting with the culture of *EL250* cells which harboured the BAC clone of interest, from 15 ml we obtained ~100 colonies from the heat-shocked cell plate after the EP, with more than 90% of the colonies containing precisely the modified BAC clone. To excise the selection cassette from the modified BACs, arabinose was added to the log phase liquid culture of the selected clone, with the correct recombinant BACs at 0.1% v/v for 1 hour at 32°C so as to sufficiently induce the *Flpe* gene in *EL250* cells. The loss of the selection cassette was monitored on the L-plate containing the antibiotic for the selection cassette. Homologous recombination events were always verified by PCR and/or electrophoresis after purification of the BAC DNA from each colony, as mentioned below.

2.2 BAC purification and evaluation

In order to obtain sufficient BAC DNA for PCR based clone selections, pulse field gel electrophoresis (PFGE) mediated evaluations and/or direct sequencing, around 3~6 ml liquid culture was inoculated for each colony. Prior to the purification of the BAC DNA, 400 μl of the saturated culture was mixed with 80 μl of 40% glycerol solution and stored at -80°C. The general protocol to purify the plasmid DNA was modified for the BAC DNA as follows: a bacterial cell pellet from the saturated culture (up to 3 ml) was collected into a 1.5 ml tube by serial centrifugations, and was suspended into 100 μl of solution I. The suspension was gently lysed by adding 200 μl of freshly prepared solution II (slowly inverting the tube four times so as to make the suspension nearly transparent) and carefully neutralised by adding 150 μl of cold solution III (gently inverting the tube twice to mix the contents evenly). After centrifugation at 15,000 rpm for 10 min at 4°C, the supernatant was overlaid on 500 μl of phenol chloroform mixture, and the content was evenly blended by gently shaking the tube three times. After centrifugation at 15,000 rpm for 5 min at room temperature, the upper-layer was transferred to a 1.5 ml tube filled with 350 μl of isopropanol, mixed and reserved at -30°C for at least 30 min. After centrifugation at 15,000 rpm for 10 min at 4°C, the pellet including the BACs and RNA was washed with 75% ethanol and dissolved into 30 μl of distilled water (DW) containing RNase at 60°C for 10min. 7.5 μl of the BAC solution can be used for visualising the DNA fragment on normal 0.6% agarose/TAE gel electrophoresis or 1% agarose/TBE PFGE. For sequencing, polyethyleneglycol-based precipitation was carried out overnight at 4°C, and the pellet was dissolved into DW to measure the optical density at OD_{260} by NanoDrop (Thermo Fisher Scientific).

To obtain good amounts of BAC DNA with high quality for trans-genesis, a CsCl density gradient ultracentrifugation-based purification method or a purification kit (NucleoBond BAC 100, Macherey-Nagel) were utilised. For CsCl purification, 1,500 ml LB was inoculated for each clone from the glycerol stock. For the kit, 250 ml of the culture was harvested and BAC DNA was purified as the manufacture's protocols recommended.

2.3 BAC trans-genesis

All the animal experiments in this study conform to Japanese governmental guidelines and have been approved by the Animal Care and Use Committee of the National Institute of Neuroscience (Projects 2007022 and 2011007). For microinjection of BAC DNA into fertilised mouse eggs, an engineered BAC clone was linearised by PI-*Sce*I, purified by ethanol precipitation and dialysed on a filter membrane (Millipore VMWP-pore size 0.05μm) to the BAC injection buffer, containing 10 mM Tris-HCl pH 7.5, 0.1 mM EDTA, 100 mM NaCl, 30 μM spermine (tetrahydrochloride; Sigma S-1141) and 70 μM spermidine (trihydrochloride; Sigma S-2501) for 2 hours. The quality of the DNA was evaluated by NanoDrop (Thermo Fisher Scientific) and the linearised BAC solution was diluted to ~2 ng/μl by the BAC injection buffer and was microinjected into pronuclei of the mouse-fertilised eggs prepared from the superovulated B6C3F1 mouse strain (SLC, Japan or Charles River, Japan). Generally, ~10 transgenic founders were obtained from ~200 eggs injected.

For the genotyping of the transgenic mouse founders or embryos, tails or yolk sacs were collected and treated with 100 μg/ml Proteinase K (Wako, Japan) at 55°C for several hours in a tail buffer containing 10 mM Tris pH7.5, 100 mM NaCl, 1 mM EDTA and 1% SDS. After heating at 95°C for 5 min, these samples were extracted once with a phenol chloroform mixture and stored at -20°C. To determine their genotypes, PCR was performed with the primer sets covering exogenous gene cassettes, such as *LacZpA*. The presence of *RP23/24* vector sequences immediately upstream or downstream of the PI-*Sce*I site was further examined by PCR so as to minimise the possibility that fortuitous large deletions fell onto the BACs which had been integrated into the chromosome.

2.4 The detection of beta-galactosidase activities or green fluorescent protein in brain slice preparations

In order to make brain slices only for the mice harbouring BAC-beta-galactosidase transgenes, we devised the rapid genotyping method for BAC transgenic pups to be finished during the morning of the day for brain dissection and slicing, as transgenic mice were always kept with the transgene being hemizygous, and were inter-crossed with B6C3F1 wild type mice (anticipating the delivery of transgenic pups at a 50% probability). Briefly, the cranial part of a mouse pup was anesthetising on ice and was placed into a chamber of the 12-well dish and filled with ice-cold Tyrode's solution. The tail was cut at the same time, and put into 100 μl of tail buffer containing 100 μg/ml of Proteinase K (see above). After finishing the collecting of the cranial parts and tails from a litter, the tail samples were incubated at 55°C for ~40 min with thorough voltexing in every 10 min. The tail samples were heated at 95°C for 5 min and extracted once with a phenol chloroform mixture. A PCR for genotyping was then performed, as described earlier (Inoue et al., 2008). While the PCR was running, the whole brain was dissected out from the cranial part and fixed in a **p**hosphate-**b**uffered **s**aline (PBS pH7.4) containing 1% paraformaldehyde, 0.1% glutalaldehyde, 2 mM $MgCl_2$, 5 mM EGTA and 0.02% Igepal CA-630 (NP-40; Sigma) for 90 min on ice. After a washing for the fixative by the washing buffer (WB: PBS containing 0.02% Igepal CA-630), these whole brain samples were kept in WB on ice. In the afternoon, we embedded only those brains confirmed to be positive for the transgene by PCR in 2% agarose/PBS, and made 550 μm slices by using a microslicer (DTK-3000, D.S.K., Kyoto). The slices were then incubated in the staining solution, containing 5 mM $K_3Fe(CN)_6$, 5 mM $K_4Fe(CN)_6 \cdot 3H_2O$, 2 mM $MgCl_2$, 0.01% sodium deoxycholate, 0.02% Igepal CA-630 and 0.1% X-gal (Wako, Japan) for several hours at 37°C with gentle agitations. The colour-detection was stopped by several washings with WB, and samples were post-fixed overnight in a PBS containing 5 mM EDTA, 1% paraformaldehyde, 0.1% glutalaldehyde and 0.02% Igepal CA-630 at 4°C. After washings with WB, the samples were finally stored at 4°C in a PBS containing 1 mM EDTA and 0.02% Igepal CA-630. For analysis, stained slice samples were put on 2% agarose/PBS plate filled with a PBS solution and flattened by a cover glass. Sample images were then captured under binocular microscope (MZ8, Leica) equipped with a CCD camera (DFC300FX, Leica) and printed out by using a video printer (SCT-CP7000, Mitsubishi, Japan).

For preparing the brain slices for the mice that harbour BAC-GAP43-EGFP transgenes, the whole brain was dissected out and illuminated under a fluorescent binocular microscope (FLIII, Leica) so as to easily select the transgenic brain. Transgenic brains were fixed in 2% paraformaldehyde/PBS for 1 hour and embedded in 2% agarose/PBS. 550 μm slices were then made by using the microslicer and photographs were taken under the microscope (FLIII, Leica) equipped with a CCD camera (DFC300FX, Leica).

2.5 Retinoic acid administration to mouse embryos and pups

To administer retinoic acid (RA; Sigma #2625 or ITSUU laboratory), the stock solution was prepared with a concentration of 30 mg/ml, and 1ml of the 1/1000 diluent (30 μg) was intraperitoneally administrated to the mother or the pups, per 1g of body weight (Smith et al. 2001).

2.6 Solutions used in the present study

LB: One litre of LB solution contains 10 g of Tryptone (BD Biosciences), 5 g of Yeast Extract (BD Biosciences), 5 g of NaCl and 4 ml of 1N NaOH. The solution is stored at room temperature after autoclave.

Solution I for BAC purification: This solution contains 50 mM glucose, 25 mM Tris-Cl (pH 8.0) and 10 mM EDTA (pH 8.0). The solution was stored at room temperature.

Solution II for BAC purification: This solution contains 1% sodium dodecyl sulphate (SDS) and 0.2N NaOH. To obtain a good amount and the purity of the BACs, the solution was freshly prepared from 10% SDS and 2N NaOH stock solutions stored at room temperature.

Solution III for BAC purification: 3M potassium acetate was dissolved into distilled water and the pH is adjusted with formic acid to 4.8. We store the solution at -20°C.

Phenol Chloroform mixture: 10mM Tris-Cl saturated phenol solution was mixed with the 0.96 volume of chloroform and 0.04 volume of isoamyl alcohol. The mixture was stored at 4°C.

PBS: One litre of 10x PBS stock contains 80 g of NaCl, 2 g of KCl, 11.5 g of Na_2HPO_4 and 2 g of KH_2PO_4. We sterilised the stock solution by autoclaving and stored it at room temperature.

TAE: One litre of 50x TAE solution contains 242 g of Tris-Cl, 57.1 ml of CH_3COOH and 100 ml of 0.5M EDTA (pH8.0).

TBE: One litre of 10x TBE solution contains 108 g of Tris-Cl, 55 g of boric acid and 40 ml of 0.5M EDTA (pH8.0).

Tyrode's solution: Five litres of Tyrode's solution contains 40.0 g of NaCl, 1.0 g of KCl, 1.0 g of $CaCl_2$, 1.05 g of $MgCl_2 \cdot 6H_2O$, 0.285 g of $NaH_2PO_4 \cdot 2H_2O$, 5.0 g of $NaHCO_3$ and 5.0 g of glucose. To avoid the possible precipitation of salts, we added these reagents in this order and sterilised by filtration and stored at 4°C.

3. Results and discussions

3.1 *Cdh6* gene expression along cortical layer IV neurons in the mouse primary somatosensory barrel field is regulated by an inter-genic region

The cerebral cortex (also called the 'neocortex' or the 'isocortex') is a mammal-specific brain region with layered cellular organisation in its radial direction (Rakic, 1988). It can further be subdivided into functional areas in its tangential direction, with each area harbouring distinct layer components, constituting fundamental units for higher brain functions that are unique to mammals (O'Leary et al., 2007). How this characteristic brain region can emerge during development as well as through evolution has been one of the important research subjects in the field of neuroscience; however, little is known about the genetic cascade required to elaborate the intricate cytoarchitecture in the cerebral cortex. In this context, classic cadherins have very unique features in their expression patterns: each cadherin subclass shows a distinct cortical layer and/or area specificity at the perinatal stages in mouse and other mammalian species, such as ferrets and humans (Krishna-K et al., 2009; Suzuki et al., 1997; Wang et al., 2009). Classic cadherins might, therefore, provide ideal genetic clues in systematically understanding the molecular mechanism of cortical development.

In the previous studies, we have focused on mouse cadherin-6 (*Cdh6*), one of classic cadherin subclasses whose expression demarcates subsets of cortical layers and/or areas (Suzuki et al., 1997; Inoue et al., 1998), and we found out that a 58 kbp long 3 prime territory to mouse *Cdh6* gene (Segment X in Inoue et al., 2008a) is required for its mRNA expression along the cortical layer IV-barrel neurons in the primary somatosensory area (S1) at postnatal day 7 (P7; Inoue et al., 2008b; our unpublished data). In the present study, we sought to further narrow down the responsible territory for the expression by systematically deleting genomic regions from the reporter modified BAC clone that can recapitulate *Cdh6* expression in S1-barrel layer IV neurons at P7. To this end, we first referred to the evolutionary conserved region (ECR) browser in which conserved genomic regions among various species are aligned and annotated (http://ecrbrowser.dcode.org). Compared with human and dog genome sequences, as many as 60 ECRs - with more than 70% similarities in a window larger than 100 bp - were found in Segment X (data not shown), and we roughly divided Segment X into three regions (regions a~c in Figure 1) by means of ECR locations. We then sought to differentially delete two of the regions in Segment X from the reporter modified BAC clone *RP23-78N21* so as to effectively narrow down the responsible territory (Asami et al., 2011; Constructs #1 and #2 in Fig. 1). We subsequently obtained three

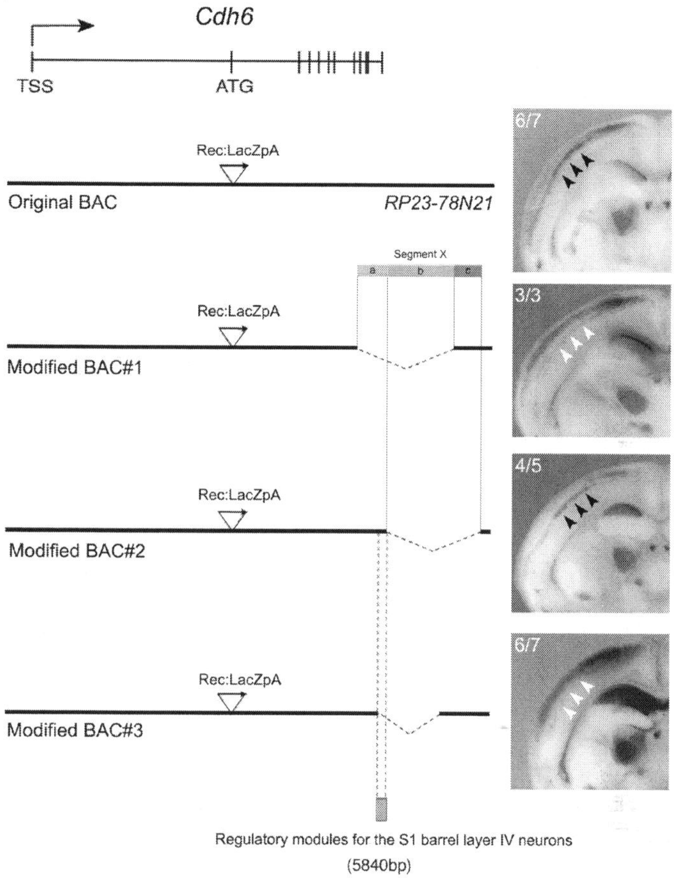

Fig. 1. An intergenic segment of mouse *Cdh6* is found to be necessary for the postnatal barrel area specific expression in the cerebral cortex.

The uppermost part of the figure indicates the genomic structure of mouse *Cdh6*, with its exons being designated by short vertical lines. ATG, translation start site; TSS, transcription start site. A BAC clone *RP23-78N21* is initially modified to harbour a beta-galactosidase gene cassette and a SV40 polyadenylation signal (*LacZpA)* in a frame to the *Cdh6* gene via homologous recombination (Rec) in a recombinogenic bacterial strain. This original BAC is further engineered by homologous recombination so as to generate deletion constructs #1~#3. Note that transgenic (Tg) mice with original BACs and construct #2 strongly recapitulate the somatosensory barrel-specific expressions (black arrow heads), while those with constructs #1 and #3 do not yield the expression at the postnatal day 7 (P7; white arrow heads). From these results, a 5840-bp territory (orange box) is determined to be a critical regulatory region for the barrel-specific expression of *Cdh6* at P7. At the upper left corner of each panel, the ratio of brain samples exhibiting reproducible reporter expression over the total number of independent transgenic mouse lines generated is noted.

stable, independent BAC transgenic (Tg) mouse founders from Construct #1 and five from Construct #2. In the former founders, none of them recapitulated the reporter expression in S1-barrel layer IV neurons (Fig. 1, white arrow heads), which are strongly marked by the original BAC trans-genesis (Fig. 1, black arrow heads). In contrast, four out of five founders reproduced the intense expression profile of S1-barrel layer IV neurons, while one of them showed no reporter expression in the brain (probably due to the positional effect of the BACs' integration site). From these results, it is strongly suggested that the most 5 prime third of Segment X (region a in Fig. 1) is responsible for *Cdh6* expression in the S1-barrel layer IV neurons at P7.

In order to further to narrow down the responsible territory, we designed the Construct #3 in which a fragment containing the most 3 prime third of region a is excluded from the original BACs. Among seven Tg founders generated from Construct #3, we could not observe the intense reporter expression of S1-barrel layer IV neurons at all (white arrow heads in Fig. 1), while we found that neurons in the other cortical layers (i.e. layers II/III) and/or areas at P7 maintained their conspicuous expression compared to the original *Cdh6*-BAC-Tg lines (Fig. 1). Taken together, we concluded that a 5,884 bp territory containing 11 ECRs is required for *Cdh6* expression in S1-barrel layer IV neurons at P7.

To our knowledge, this is the first time that a distinct gene regulatory fragment for a defined layer and/or area has been observed, suggesting that separable genetic programs may serve the pattering of each cortical layer and/or area during development. The further characterisation of the gene regulatory elements that directly interact with the 5,884 bp territory would be an important next step in understanding how the cortical layer and/or area identity is strictly determined during development. In this connection, it is noticeable that the 5,884 bp territory contains many of the transcription factor binding motifs, such as *RORbeta*, whose expression is already known to be restricted to defined sets of cortical layers and/or areas (Dye et al., 2011; Hirokawa et al., 2008; Nakagawa & O'leary, 2003). Since recent reports suggest that *RORbeta* harbours an instructive role in elaborating barrel cytoarchitecture and/or circuitries (Jabaudon et al., 2011), it would be of great interest to rigorously evaluate how these transcription factors are involved in establishing *Cdh6* expression along S1-barrel layer IV neurons at P7 which might have functional significance in driving cell segregations to form and/or maintain the barrel cytoarchitecture.

3.2 Retinoic acid (RA) related cell signalling machineries might be involved in the cortical barrel patterning

RA signalling regulates many of the morphogenetic events at the early embryonic stage, such as A-P axis formation, the establishment of L-R asymmetry and so on (Kiecker & Lumsden, 2005; Niederreither & Dolle, 2008). It has recently been suggested that RA might also have physiological roles in cortical development at the later embryonic stages (Siegenthaler et al., 2006; Smith et al., 2001). We thus next tried to examine the possible roles of RA related cell signalling machinery in the cortical area pattering during development. By using the same BAC clone *RP23-78N21*, we replaced the *Cdh6*-ATG exon to an expression cassette for the membrane-bound form of enhanced green fluorescent protein (GAP43-EGFP) by means of homologous recombination and generated transgenic

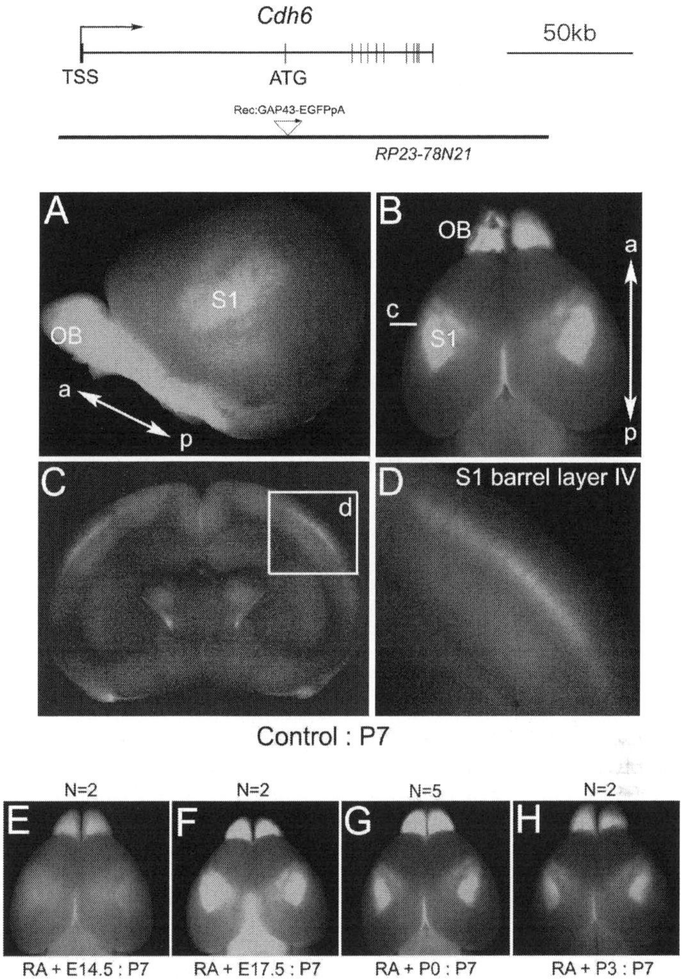

Fig. 2. Possible involvement of retinoic acid (RA) related cell signalling machineries in the cortical barrel patterning.
A BAC clone *RP23-78N21* is modified to harbour a membrane-bound form of EGFP cassette and SV40 polyadenylation signal (GAP43-EGFP-pA) in frame to the *Cdh6* gene via homologous recombination (Rec) in a recombinogenic bacterial strain (upper most part of the figure). (A-D) A Tg mouse with EGFP modified BACs recapitulated *Cdh6* expression at P7. OB, olfactory bulb; S1 primary somatosensory area. The boxed area d in panel C is magnified in panel D. Note that Layer IV barrels in S1 are illuminated in this Tg mouse line. (E-H) Effects of RA on barrel development. RA is intraperitoneally injected into the mother (E, F) or the pups (G, H) and the EGFP expression profile is evaluated at P7. Note that this is only the case with E14.5 injections, which affect the formation of barrel cytoarchitecture, highlighting the role of RA in early neocortical development.

mouse founders (Inoue et al., 2009). In the Tg cerebral cortex, we found that GFP expression shows exactly the same patterns as with the original LacZ-Tg mice (Fig 1 and 2). In particular, this GFP-BAC-Tg mouse line illuminated the S1 barrel structure of the whole mount brain preparations, allowing us to easily image the S1-barrel territory whose identification generally requires specific histological staining processes, such as the CO staining method.

We then administrated RA to the mothers or pups for GFP-BAC-Tg, with the concentration reported to induce abnormality in the cortex (Smith et al., 2001) and evaluated how RA affects the GFP expression patterns at P7. As a result of this, no drastic change was observed for the cortical barrel patterning when RA was administrated later than embryonic day 17.5 (E17.5). However, RA administration at E14.5 resulted in massive perturbation of S1 barrel patterning, with ambiguous area boundaries illuminated by GFP expression. Noticeably, the intensity of the GFP expression appeared to be decreased due to the qualitative and quantitative differences among *Cdh6::GFP* positive cortical cells and/or thalamocortical axon terminals. These results imply that the role of RA in regulating *Cdh6* expression and/or cortical area pattering is just limited to those embryonic stages earlier than E14.5.

It is now widely accepted that cortical area patterning begins as early as mouse E12.5 when the counter-gradient of the transcription factors Pax6 and Emx2 is established in the cortical ventricular zones (Bishop et al., 2000; Hamasaki et al., 2004). This gradation pattern, as generated by such secreted molecules as *Fgf8*, is shown to be the basis of cortical arealisation yet other transcriptional factors, such as *Coup-TFI,* could regulate the area-specific differentiation of distinct subtypes of cortical neurons independently of *Fgf8-Pax6/Emx2* gene functions (Armentano et al., 2007; Fukuchi-Shimogori & Grove, 2001, 2003). Our results, together with a previous series of studies, thus suggest that RA accumulated earlier than E14.5 might play a role in cortical arealisation by affecting the production, migration, positioning and/or circuit formation of the cortical S1 barrel layer IV neurons that eventually express *Cdh6*. The next critical step would be to examine whether RA-related signalling could be interactive with the 5,884 bp territory identified in this study that contains *RORbeta* related transcription factor binding motifs.

3.3 BACs in the field of neuroscience research

In the present study, BAC-based methodology enabled us to systematically evaluate intricate genetic machineries in the mouse brain, highlighting the value of BAC usage in the field of neuroscience research. Recently, others have also developed many useful BAC-based strategies, and here we discuss the advantages and/or future potential of some of these strategies in the field of neuroscience research.

First, we have realised that BACs must provide a useful and ideal basis in approaching the complex gene regulatory machinery that elaborates the nervous system. For instance, a difficulty involved in studying the vertebrate nervous system has lain in its complexity and, when we try to rigorously dissect its specialised structure and function, it should be an essential step in discriminating a defined group of cells among numerous neurons and glial cells. As has been demonstrated by the present study, homologous recombination-based systematic deletions from BAC clones combined with efficient mouse trans-genesis now allows for the quick identification of *cis*-regulatory elements from the huge *Cdh6* gene locus, which contrasts with the conventional methodology where hundreds of conserved genomic regions must be evaluated one-by-one by means of plasmid-based methods (See discussions in Asami et al., 2011). A problem with this recombination-based method might lie in its limit for BAC clone usage, since the BAC clone for recombination must always include an ATG translation initiation codon for the gene of interest to achieve in-frame integration of the reporter cassette. This being the case, *cis*-regulatory elements located far outside of the ATG-containing BAC clone cannot also be treated by this method. However, if combined with the transposon-based BAC modification method, one can reliably monitor the gene transcriptional activity of any BAC clones, regardless of their gene/ATG-exon coverage (Asami et al., 2011). Hence, a BAC-based strategy would greatly help the understanding of the upstream/downstream relationships among thousands of genes that are preferentially expressed in the nervous system, revealing the entire genetic programmes for building up the nervous system. Noticeably, many human single nucleotide polymorphisms (SNPs) that are tightly linked to genetic disorders have been identified in the intergenic regions (Wang et al., 2009). These SNPs are thought to play roles in regulating gene expression and/or chromosomal structures, yet the methodologies that can reliably detect their functional significance are limited in number at the present time. In this context, our BAC-based methodology could immediately serve as a steadfast platform in approaching such critical research subjects (Asami et al., 2011).

Secondly, we have demonstrated that BAC-based methodology is very useful for the stable and efficient labelling of distinct cells-types during neural development. Since fluorescent proteins - such as GFP - can easily visualise cell morphology, they have been broadly used, from basic molecular-cell biology to biomedical studies. In the field of neuroscience research, it remains a fundamental issue to thoroughly identify the original location of neurons and their partners among the billions of neurons, since they often make contact with other neurons located far away from the soma. For this purpose, the use of a membrane localisation signal, such as GAP43 tagged GFP, would clearly visualise axons, while nuclear localisation signal-based reporter introduction would precisely identify the soma location. Recently, a rabies virus mediated single synaptic transfer event was applied so as to enable systematic labelling of both the starter neurons and their primary partners in connections (Miyamichi et al., 2011). However, in order to restrict the number of neurons labelled by this method in the nervous system, it is most critical to identify enhancers/promoters that confine gene expression in the limited population of cells amongst the tens of thousands of neurons. In this context, BACs do provide an ideal resource for such an analysis because we can now easily engineer a given BAC clone to include sets of enhancers/promoters for obtaining restricted gene expression profiles, and the GENSAT project indeed generated hundreds of BAC transgenic mouse lines that differentially illuminate specific sets of cells in the nervous system (Gong et al., 2003; Gong et al., 2010). If a GFP tagged L10a ribosomal protein is expressed by BAC trans-genesis, the mRNA expressed in the specific set of cells should be illuminated and can further be isolated by the fluorescent activated cell-sorting system so as to profile their molecular characteristics (Heiman et al., 2008).

Thirdly, a BAC-based methodology is highly expected to open a new window into the study of unknown processes for brain development and/or the functional dynamics of neural circuitries. For instance, in multicellular model organisms, such as the mouse, the fly and the nematode, loss of function studies are currently the gold standard for revealing given gene functions, contributing to the revelation of genetic programmes at the early developmental stages. There had been, however, a problem in that a simple loss of function analysis sometimes results in early embryonic lethality, preventing researchers from evaluating the gene functions in mature organs, such as brains. To circumvent this situation, a conditional gene knock-out strategy was established so that a given gene function is abolished at a defined time and place by genetically introducing the enhancer-driven site-specific recombinase *Cre* and its recognition sites LoxP sequences into the gene locus of interest. Considering their extensive coverage of various enhancers/promoters in the genome of multicellular model organisms, BAC clones should serve as the perfect starting points in the establishment of useful driver transgenic animals for conditional knock-out studies. Additionally, if the expression cassette for *Cre* and the estrogen receptor T2 variant fusion protein (*CreERT2*; Feil et al., 1997) is integrated into a proper BAC clone to generate Tg animals in which *CreERT2* proteins are expressed among a limited group of cells, one can precisely control the timing to generate gene mutant cells by merely administrating tamoxifen, which allows selective *CreERT2* localisation into the cell nuclei so as to excise the gene of interest by recombination. *CreERT2*-Tg animals might further be suitable for genetic cell-lineage tracing. In the mouse system, this can generally be achieved by using the reporter mouse lines, such as *Rosa26R*, in which *LacZ* reporter expression is suppressed by the intercalation of a stopper put in between the LoxP sequences. When this reporter line is mated with the *CreERT2*-Tg mouse line, the stopper is excised only with the administration of tamoxifen and, thereafter, a limited population of cells will be genetically and permanently marked by the reporter expression. Indeed, we have generated the *Cdh6::CreERT2*-BAC-Tg mouse to clarify the relationship between the *Cdh6* gene expression boundary at the early cortical plate and the mature areal boundary, and have found a rigid correlation (Terakawa et al., manuscript submitted). Useful *CreERT2*-Tg mouse lines could be further be mated with the recently created Brainbow mouse Tg line, logically allowing the genetic labelling of individual cells in the nervous system by different fluorescent colour combinations (Livet et al., 2007). Such spatio-temporally regulated labelling of cells must aid in unveiling the functional dynamics of the nervous system in higher vertebrates with complex cellular organisation.

To finally address the fundamental question as to how the elaborated neural circuitries work in the *in vivo* context, the BAC-based introduction of optogenetic probes, such as channel rhodopsin and halorhodopsin, into specific sets of neurons might make it possible to selectively switch on and off neuronal activities within the regions that receive the relevant light stimuli (O'Connor et al., 2009; Zhang et al., 2007). Since the individual optogenetic probe harbours different wavelengths' selectivity, defined sets of neuronal and/or muscle activities can be manipulated by simply applying combinatorial light stimuli to the Tg animals. This technology therefore speeds the detailing of which circuitries are actually responsible for a given behaviour and/or the process for learning and memory, exemplifying how BAC-related experimental methods can be applicable to wide range of research in the field of neuroscience.

4. Conclusions

Taking advantage of systematic BAC modification methodologies via homologous recombination and/or transposon tagging in bacterial cells, as well as efficient BAC transgenic strategies in various multicellular organisms, such as mice, it is now possible to mark and manipulate a given gene function amongst restricted cell groups in the nervous system at will. Given that BACs constitute the minimal components of various whole genome-sequencing projects, BAC-based technology would significantly facilitate the detail of entire genetic programmes that elaborate the complex structure and function of the nervous system, including our own. Such detailed information would greatly help the appreciation of the intricate principles of neural evolution and development, encoding, processing and/or pathogenesis.

5. Acknowledgments

We thank Dr. Robb Krumlauf for pBGZ40 vector and Dr. Neal G. Copeland for the recombineering related materials including *EL250* bacterial stain. We also acknowledge Drs. Takayuki Sota, Shinichi Kohsaka, Keiji Wada, Mikio Hoshino and their lab members for discussions and encouragements. This work was supported by grants from Takeda Science Foundation, The Nakatomi Foundation and Research Foundation ITSUU Laboratory, Program for Promotion of Fundamental Studies in Health Sciences of the National Institute of Biomedical Innovation (05-32), and a Grant-in-Aid for Scientific Research from JSPS (#21500333) to T. Inoue.

6. References

Armentano M, Chou SJ, Tomassy GS, Leingärtner A, O'Leary DD, Studer M. 2007. COUP-TFI regulates the balance of cortical patterning between frontal/motor and sensory areas. Nat Neurosci. 10:1277-1286.

Asami J, Inoue YU, Terakawa YW, Egusa SF, Inoue T. 2011. Bacterial artificial chromosomes as analytical basis for gene transcriptional machineries. Transgenic Res. 20:913-924.

Bishop KM, Goudreau G, O'Leary DD. 2000. Regulation of area identity in the mammalian neocortex by Emx2 and Pax6. Science. 288:344-349.

Copeland NG, Jenkins NA, Court DL. 2001. Recombineering: a powerful new tool for mouse functional genomics. Nat Rev Genet. 2:769-779.

Dye CA, El Shawa H, Huffman KJ. 2011. A lifespan analysis of intraneocortical connections and gene expression in the mouse II. Cereb Cortex. 21:1331-1350.

Feil R, Wagner J, Metzger D, Chambon P. 1997. Regulation of Cre recombinase activity by mutated estrogen receptor ligand-binding domains. Biochem Biophys Res Commun. 237:752-757.

Fujimori T, Miyatani S, Takeichi M. 1990. Ectopic expression of N-cadherin perturbs histogenesis in Xenopus embryos. Development. 110:97-104.

Fukuchi-Shimogori T, Grove EA. 2001. Neocortex patterning by the secreted signalling molecule FGF8. Science. 294:1071-1074.

Fukuchi-Shimogori T, Grove EA. 2003. Emx2 patterns the neocortex by regulating FGF positional signalling. Nat Neurosci. 6:825-831.

Gong S, Kus L, Heintz N. 2010. Rapid bacterial artificial chromosome modification for large-scale mouse transgenesis. Nat Protoc. 5:1678-1696.

Gong S, Zheng C, Doughty ML, Losos K, Didkovsky N, Schambra UB, Nowak NJ, Joyner A, Leblanc G, Hatten ME, Heintz N. 2003. A gene expression atlas of the central nervous system based on bacterial artificial chromosomes. Nature. 425:917-925.

Gumbiner BM. 2005. Regulation of cadherin-mediated adhesion in morphogenesis. Nat Rev Mol Cell Biol. 6:622-634.

Hamasaki T, Leingärtner A, Ringstedt T, O'Leary DD. 2004. EMX2 regulates sizes and positioning of the primary sensory and motor areas in neocortex by direct specification of cortical progenitors. Neuron. 43:359-372.

Heiman M, Schaefer A, Gong S, Peterson JD, Day M, Ramsey KE, Suárez-Fariñas M, Schwarz C, Stephan DA, Surmeier DJ, Greengard P, Heintz N. 2008. A translational profiling approach for the molecular characterization of CNS cell types. Cell. 135:738-748.

Hirokawa J, Watakabe A, Ohsawa S, Yamamori T. 2008. Analysis of area-specific expression patterns of RORbeta, ER81 and Nurr1 mRNAs in rat neocortex by double in situ hybridization and cortical box method. PLoS One. 3:e3266.

Inoue T, Inoue YU, Asami J, Izumi H, Nakamura S, Krumlauf R. 2008a. Analysis of mouse Cdh6 gene regulation by transgenesis of modified bacterial artificial chromosomes. Dev Biol. 315:506-520.

Inoue T, Tanaka T, Suzuki SC, Takeichi M. 1998. Cadherin-6 in the developing mouse brain: expression along restricted connection systems and synaptic localization suggest a potential role in neuronal circuitry. Dev Dyn. 211:338-351.

Inoue YU, Asami J, Inoue T. 2008b. Cadherin-6 gene regulatory patterns in the postnatal mouse brain. Mol Cell Neurosci. 39:95-104.

Inoue YU, Asami J, Inoue T. 2009. Genetic labelling of mouse rhombomeres by Cadherin-6::EGFP-BAC transgenesis underscores the role of cadherins in hindbrain compartmentalization. Neurosci Res. 63:2-9.

Jabaudon D, J Shnider S, J Tischfield D, J Galazo M, Macklis JD. 2011. ROR{beta} Induces Barrel-like Neuronal Clusters in the Developing Neocortex. Cereb Cortex.

Kiecker C, Lumsden A. 2005. Compartments and their boundaries in vertebrate brain development. Nat Rev Neurosci. 6:553-564.

Krishna-K, Nuernberger M, Weth F, Redies C. 2009. Layer-specific expression of multiple cadherins in the developing visual cortex (V1) of the ferret. Cereb Cortex. 19:388-401.

Lee EC, Yu D, Martinez de Velasco J, Tessarollo L, Swing DA, Court DL, Jenkins NA, Copeland NG. 2001. A highly efficient Escherichia coli-based chromosome engineering system adapted for recombinogenic targeting and subcloning of BAC DNA. Genomics. 73:56-65.

Livet J, Weissman TA, Kang H, Draft RW, Lu J, Bennis RA, Sanes JR, Lichtman JW. 2007. Transgenic strategies for combinatorial expression of fluorescent proteins in the nervous system. Nature. 450:56-62.

Miyamichi K, Amat F, Moussavi F, Wang C, Wickersham I, Wall NR, Taniguchi H, Tasic B, Huang ZJ, He Z, Callaway EM, Horowitz MA, Luo L. 2011. Cortical representations of olfactory input by trans-synaptic tracing. Nature. 472:191-196.

Nakagawa S, Takeichi M. 1998. Neural crest emigration from the neural tube depends on regulated cadherin expression. Development. 125:2963-2971.

Nakagawa Y, O'Leary DD. 2003. Dynamic patterned expression of orphan nuclear receptor genes RORalpha and RORbeta in developing mouse forebrain. Dev Neurosci. 25:234-244.

Niederreither K, Dollé P. 2008. Retinoic acid in development: towards an integrated view. Nat Rev Genet. 9:541-553.

Nose A, Nagafuchi A, Takeichi M. 1988. Expressed recombinant cadherins mediate cell sorting in model systems. Cell. 54:993-1001.

O'Connor DH, Huber D, Svoboda K. 2009. Reverse engineering the mouse brain. Nature. 461:923-929.

O'Leary DD, Chou SJ, Sahara S. 2007. Area patterning of the mammalian cortex. Neuron. 56:252-269.

Rakic P. 1988. Specification of cerebral cortical areas. Science. 241:170-176.

Redies C. 2000. Cadherins in the central nervous system. Prog Neurobiol. 61:611-648.

Siegenthaler JA, Ashique AM, Zarbalis K, Patterson KP, Hecht JH, Kane MA, Folias AE, Choe Y, May SR, Kume T, Napoli JL, Peterson AS, Pleasure SJ. 2009. Retinoic acid from the meninges regulates cortical neuron generation. Cell. 139:597-609.

Smith D, Wagner E, Koul O, McCaffery P, Dräger UC. 2001. Retinoic acid synthesis for the developing telencephalon. Cereb Cortex. 11:894-905.

Steinberg MS, Takeichi M. 1994. Experimental specification of cell sorting, tissue spreading, and specific spatial patterning by quantitative differences in cadherin expression. Proc Natl Acad Sci U S A. 91:206-209.

Suzuki SC, Inoue T, Kimura Y, Tanaka T, Takeichi M. 1997. Neuronal circuits are subdivided by differential expression of type-II classic cadherins in postnatal mouse brains. Mol Cell Neurosci. 9:433-447.

Takeichi M. 1995. Morphogenetic roles of classic cadherins. Curr Opin Cell Biol. 7:619-627.

Takeichi M, Abe K. 2005. Synaptic contact dynamics controlled by cadherin and catenins. Trends Cell Biol. 15:216-221.

Wang K, Zhang H, Ma D, Bucan M, Glessner JT, Abrahams BS, Salyakina D, Imielinski M, Bradfield JP, Sleiman PM, Kim CE, Hou C, Frackelton E, Chiavacci R, Takahashi N, Sakurai T, Rappaport E, Lajonchere CM, Munson J, Estes A, Korvatska O, Piven J, Sonnenblick LI, Alvarez Retuerto AI, Herman EI, Dong H, Hutman T, Sigman M, Ozonoff S, Klin A, Owley T, Sweeney JA, Brune CW, Cantor RM, Bernier R, Gilbert JR, Cuccaro ML, McMahon WM, Miller J, State MW, Wassink TH, Coon H, Levy SE, Schultz RT, Nurnberger JI, Haines JL, Sutcliffe JS, Cook EH, Minshew NJ, Buxbaum JD, Dawson G, Grant SF, Geschwind DH, Pericak-Vance MA, Schellenberg GD, Hakonarson H. 2009. Common genetic variants on 5p14.1 associate with autism spectrum disorders. Nature. 459:528-533.

Zhang F, Aravanis AM, Adamantidis A, de Lecea L, Deisseroth K. 2007. Circuit-breakers: optical technologies for probing neural signals and systems. Nat Rev Neurosci. 8:577-581.

8

Production of Multi-Purpose BAC Clones in the Novel *Bacillus subtilis* Based Host Systems

Shinya Kaneko[1] and Mitsuhiro Itaya[2]
[1]*Graduate School of Bioscience and Biotechnology, Department of Life Science*
Tokyo Institute of Technology, Yokohama-shi Kanagawa
[2]*Institute for Advanced Biosciences, Keio University, Yamagata*
Japan

1. Introduction

1.1 A brief history of BACs

The BAC vector that replicates in *Escherichia coli* K-12 was first introduced in 1992 by Shizuya et al. More detailed information on current BAC applications may be referred to other chapters of this book. Because BACs can accommodate much longer DNA stretches than the plasmids for *E. coli* available at that time, as large as >100 kbp, they were suitable for the preparation of DNA libraries covering entire cellular genomes regardless of their original size (Fig. 1a). The initial goal for BACs was to provide materials for whole genome-sequencing projects and

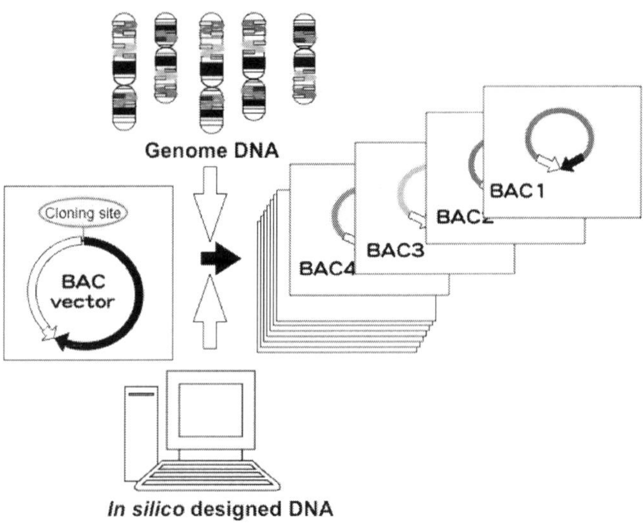

Fig. 1a. BAC library for the contiguous genomic DNA. The ability to carry large DNA (>100kb) is useful for not only the direct analysis of genomic regions in reverse genetics, but also for the *de novo* synthesis of genomic DNA in the field of synthetic biology.

their contribution to long-range sequence determinations has been demonstrated (Frengen et al., 1999; Osoegawa et al., 2001). Cutting-edge technologies that facilitate direct genome sequencing have dramatically reduced the need for BAC libraries as a sequencing resource.

1.2 BAC applications

The *de novo* synthesis of small DNA fragments has become routine although it remains expensive (Gibson et al., 2008, 2010; Itaya et al., 2008; Itaya, 2010). In that process, commercial enterprises customize desired DNA fragments based on nucleotide sequence information provided by the end-user. Larger DNA, equivalent to lengths clonable in BACs, can be prepared routinely by assembling small DNA fragments (Gibson 2011; Itaya & Tsuge, 2011). Reverse genetics methods applied in studies involving cultured cells and model animals are increasingly important in mutation research (Yang et al., 1997; Hardy et al., 2010). DNA can now be obtained by *de novo* synthesis using designed sequences or by flexible engineering of cloned DNA in BACs. Although BACs can accommodate DNA fragments longer than 100 kbp, the intrinsic physicochemical features of long-stretched polymer molecules render them fragile and their handling difficult. Due to the shearing force in liquids, DNA fragments easily break into small pieces and nuclease contamination may be introduced in the course of biochemical isolation procedures. Therefore care must be taken in the isolation and purification of large DNAs (Kaneko et al., 2005) and appropriate host cell systems are needed to nurture and protect the fragile DNA fragments to facilitate the preparation of undamaged DNA samples regardless of their size.

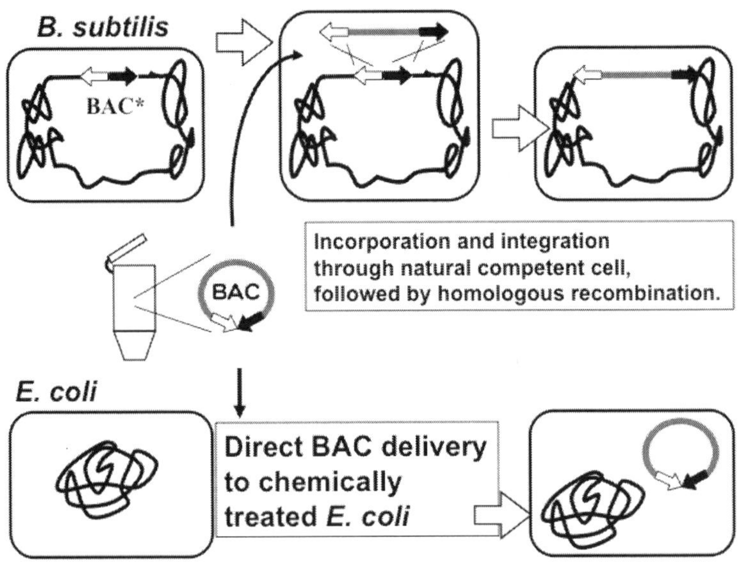

Fig. 1b. *B. subtilis* as a new BAC dealer.
The integration of BAC inserts into the genome of *B. subtilis* is a starting point for subsequent manipulations. The BAC vector region (**BAC***) preinstalled in the host genome provides the cloning site for guest BAC clones via homologous recombination (identified by X).

1.3 *E. coli* - An original host and an engineering system for BACs: Size limits for stable cloning

Of particular import with respect to the direct application of the BACs carrying genes of interest is their modification to render them useful for in-depth reverse genetics research. Some examples on current application and molecular engineering of BAC vectors can be found in other chapters of this book. Currently-available systems for handling large DNA in *E. coli* make it possible to carry out pin-point engineering on DNA carried in BACs. The BAC clone is initially maintained in autonomously-replicating plasmid form in *E. coli*. The prepared BACs are then used for delivery to an *E. coli* BAC-engineering system (Fig. 1b). Genetic manipulations on BACs are summarized in Fig. 2. The maintenance of BACs in *E. coli* requires selection by antibiotics during all stages, as is the case with other plasmid vectors.

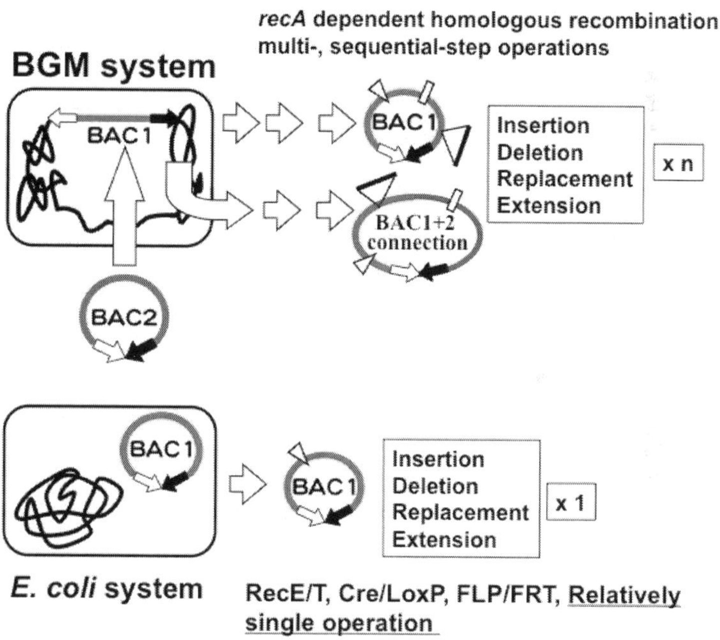

Fig. 2. The *B. subtilis* Genome (BGM) system.
Multiple manipulations of large BAC-DNA fragments are possible in BGM; in the *E. coli* system only single manipulations can be performed. The stability of DNA attributable to the one-copy state of the *B. subtilis* genome assures the maneuverability of *recA*-dependent homologous recombination. Maneuvers can be repeated unlimited times.

The minimal requirements to ensure the stability of BACs before and after engineering operations in *E. coli* are shown in Fig. 2. There are size limitations for the stable handling of BACs in *E. coli;* it has been suggested that the largest clonable size does not exceed 500 kbp. To date not even the existing minimal genome 585-kbp circular one of *Mycoplasma genitalium* has been successfully cloned in *E. coli* (Gibson et al., 2008, 2010).

2. The emerging *B. subtilis* host

B. subtilis has emerged as an appropriate host to supplement the use of BACs from *E. coli* (Itaya, 2009). Our favorite cloning host, *B. subtilis* 168, yet unfamiliar to many researchers, is a Gram-positive firmicute bacterium. It grows as rapidly and in the same media as *E. coli*. Consequently, many protocols used with *E. coli* can be used with *B. subtilis*. However, features inherent in *B. subtilis* facilitate natural transformation, in addition, the bacterium forms endospores that survive for extraordinarily long periods. *E. coli* lacks these features.

2.1 What is *B. subtilis?* - A more detailed explanation

We will place special emphasis on the use of natural transformation in the engineering of BACs (Fig. 1b). The general homologous recombination associated with *B. subtilis* transformation is distinct from the induced recombination adopted by most *E. coli* DNA engineering systems. The *B. subtilis* 168 strain domesticated in the laboratory was isolated as a recipient that facilitates DNA-mediated transformation (Spizizen, 1958). The subsequent elucidation of its detailed molecular mechanisms revealed that competent *B. subtilis* cells can incorporate extracellular DNA into their cytoplasm via the process of natural transformation (Kidane & Graumann, 2005). It is possible to conduct genetic crosses in *B. subtilis* and like *E. coli* K-12, it has been subjected to extensive biochemical and genetic analyses. Although *E. coli* was thought to accept extracellular DNA via a chemical transformation process, this type of transformation was limited to the plasmid delivery method (Mandel & Higa, 1970; Hanahan, 1983). The cloning in and the propagation of BACs in *E. coli* are subsumed under the term DNA delivery. The process of DNA delivery through the cell membrane of *E. coli* involves elaborate physical and chemical treatments of the host (Fig. 1b). In contrast, the transformation system of *B. subtilis* applies to plasmid delivery as well as genomic gene engineering. Competent *B. subtilis* cells positively grab extracellular DNA and pull the fragment(s) into their cytoplasm in single-strand form. The molecular mechanisms, the processing of double-stranded DNA from the cell surface and its conversion into single-strand DNA for entry through the cellular membrane, are carried out in a concerted manner by several proteins encoded by *B. subtilis* genes (Kovács et al., 2009). This process is too complex to be explained here; in short, *B. subtilis* does, while *E. coli* does not possess a set of genes to conduct natural transformation.

As incorporated BACs cannot replicate in *B. subtilis* in plasmid form because there is no replication origin sequence for this host, BACs must enter cellular recombination pathways initiated by the existing *recA* protein (Kidane & Graumann, 2005). If homologous sequences are present in the *B. subtilis* chromosome (BAC* in Fig. 1b), the incoming BAC is integrated into the homologous region. Such homologous recombination-mediated transformation yields the integrated form at high frequency in the *B. subtilis* genome (Fig. 1b). This well-known natural transformation system for the *B. subtilis* 168 strain has been used in genetic research targeting the original genome. Two striking results were the derivation of a 33-fold mutant of *B. subtilis* by repeating the transformation 33 times to introduce mutations at 33 chromosomal loci (Itaya & Tanaka, 1991) and the reduction of the genome to 75% of its original size by the consecutive deletion of genes unaffected by growth (Ara et al., 2007). Fortuitously *B. subtilis* 168 possesses no original/cognate plasmid.

2.2 Why is the *B. subtilis* host advantageous?

A homologous sequence is the sole requirement for extra DNA-engineering in *B. subtilis*. The transformation process shown in Fig. 2 can be repeated and the number of repetitions is practically unlimited. Consequently, due to its ability to repeat transformation, the assembly of large DNA fragments in the *B. subtilis* genome is possible. Typically, if DNA fragments with partial overlaps are prepared, repetitive integration by using the overlapping regions allows the reconstruction of the original DNA in the *B. subtilis* genome (Itaya & Tsuge, 2011). Even before its whole genome sequence determination in 1997, *B. subtilis* became a workhorse for the cloning and manipulations particularly of giant DNAs that cannot be handled by *E. coli*. As *B. subtilis* forms endospores that manifest significant resistance to vacuuming, dryness, and radiation, it has become a reservoir for giant DNA maintained at room temperature. Additional details will be presented in section 3-4.

2.3 Why the *B. subtilis* host - An unprecedented genome vector

Despite conceptual differences between *E. coli* and *B. subtilis* with respect to the cloning of DNA, *B. subtilis* is highly advantageous because it allows the use of small DNAs prepared by routine molecular cloning in *E. coli*. A high workload can be handled by *B. subtilis* via the integration of DNA fragments cloned once in an *E. coli* pBR322 plasmid or BAC. The plasmid transfer from *E. coli* to the *B. subtilis* genome shown in Fig. 1b renders the *B. subtilis* genome a big cloning vector. The transfer of *E. coli*-borne plasmids to the *B. subtilis* genome is advantageous because after the DNA is stably integrated into the *B. subtilis* genome, it becomes part of its genome. A wide range of genetic manipulations is now possible in *B. subtilis* regardless of the origin of the integrated DNA. The number and effectiveness of tools available for genetic manipulations in *B. subtilis* far exceed the tools available in the *E. coli* K-12 system. They include the faithful and stable insertion and deletion of any DNA at designated loci. Another advantage of *B. subtilis* is its extraordinary stability, this eliminates the selection pressure to maintain DNA. This is in sharp contrast with the stringent requirements for the maintenance of plasmids in *E. coli*. During growth, plasmids that replicate independent of the host genome segregate out without selection pressure (Fig. 1b and Fig. 2). Indeed, we have demonstrated that *B. subtilis* stably accommodated DNAs far larger than those covered by *E. coli* BACs. The largest size was up to 3,500 kb (Itaya et al., 2005). We think that the stability of DNA integrated into the *B. subtilis* genome is attributable to the presence of a single-copy genome per bacterial cell. In subsequent sections we focus on the handling of BACs transferred to the *B. subtilis* genome and their recovery. This vector system can handle DNA manifesting sequence variations such as short- (mouse genome) or long repeats (IR), and different GC contents.

3. Type of engineered BACs in the *B. subtilis* genome (BGM) vector

Hereafter, DNA introduced into the *B. subtilis* host will be called guest DNA. Guest DNA cloned in the host becomes integrated into the host genome. The single-copy genome of each *B. subtilis* cell can accommodate chloroplast and mitochondrial guest genomes (Itaya et al., 2008) and the genome of the bacterium *Synechocystis* PCC6803 (Itaya et al., 2005). *B. subtilis* strains developed as hosts for these DNAs are called BGM vectors, an acronym for *Bacillus* GenoMe vector. The wild-type strain *B. subtilis* 168 was the first BGM vector to host one pBR322 sequence (Itaya, 1993). This sequence was successfully introduced at different genomic loci (Itaya et al., 2005). The integrated pBR322, a 4.3 kbp *E. coli* plasmid, served as a common cloning locus in a manner reminiscent of the integration by homologous recombination illustrated in Fig 1b. Before the creation of the BAC vector, pBR322 and its derivatives were widely used for various gene-cloning experiments as they offered several advantages, e.g. a small size, a medium-sized copy number, and an ability to carry DNA up to 30 kbp. DNA cloned in pBR322 via the *E. coli* molecular cloning system immediately became guest DNA in the pBR322-based BGM. Integration required only two homologous sequences and appropriate selection markers for the bacterium. The DNA flow from *E. coli* BACs to the BGM vector is shown in Fig. 1b; it is similar to the pBR322-based system but requires major modifications.

3.1 Direct transfer of guest BACs

As commercially- or laboratory-prepared BAC clones carried no antibiotic resistance markers for *B. subtilis*, the first BAC-BGM required the pre-installation of a counter-selection system to stimulate the integration process (see Fig. 3). In our initial experiments on the integration of mouse genomic DNA carried by BACs (Kaneko et al., 2003, 2005, 2009; Itaya et al., 2000), we observed no structural disorder during the integration process despite the short repeats generally present in the mouse genome (Itaya et al., 2000; Kaneko et al., 2003, 2009). In addition, the BGM stably carried the 25-kbp-long inverted repeats present in the rice chloroplast genome (Itaya et al., 2008). Consequently, we thought that the BGM could accommodate not only very large-sized DNA but also a wide range of sequence variations from other genomes.

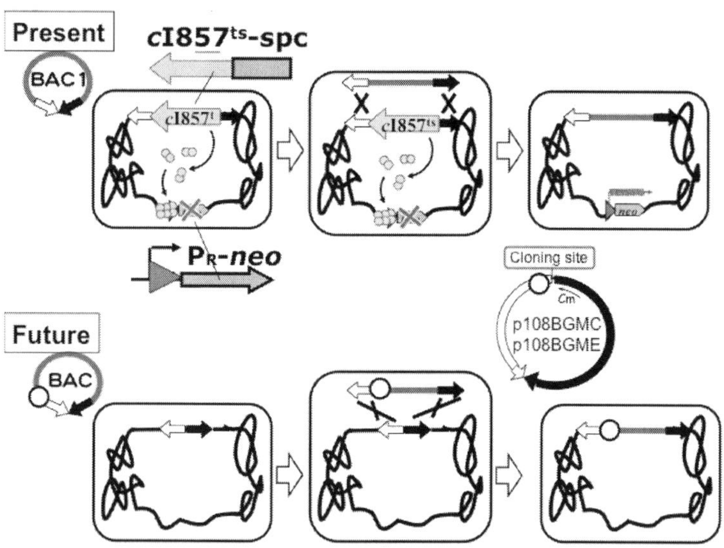

Fig. 3. Cloning of BACs carrying a non-marker for *B. subtilis*.
Top: The present counter selection system is shown. The *c*I repressor gene and the neomycin-resistance gene under the Pr prompter result in the positive selection of marker-less BACs for integration.
Bottom: BAC clones in the new BAC vectors, p108BGMC or p108BGME, carrying an antibiotics marker for *B. subtilis* can be cloned directly in BAC-BGM.

After BAC clones were regularly used both in *E. coli* and *B. subtilis*, the next step was to elaborate the engineering/manipulation of the DNA. Figure 4 illustrates the design and modification of nucleotide sequences inside the guest DNA. Examples are detailed below; various size ranges in section 3-2, connecting two overlapping BAC clones in 3-3, applying genome techniques developed for *B. subtilis* in 3-4, and the unique preservation of designed BACs in BGM for prolonged storage in the absence of special facilities in 3-5.

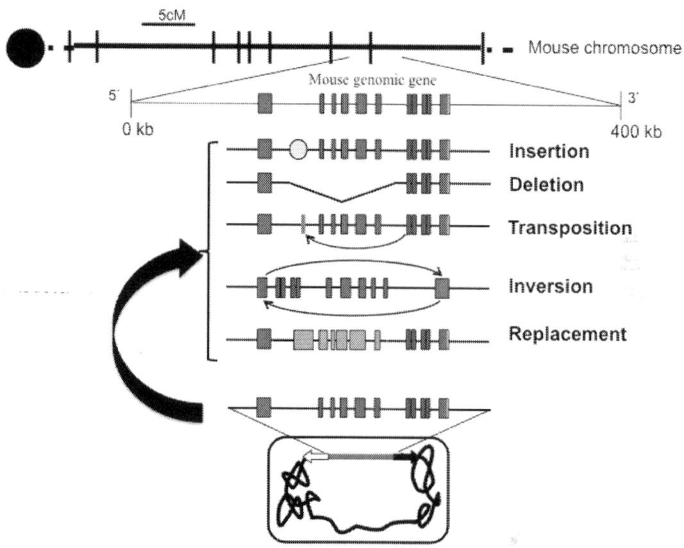

Fig. 4. Manipulation of the guest DNA.
A giant mouse genomic region integrated, for example, into BGM can be re-designed to the indicated structures by relying on the high fidelity of homologous recombination in *B. subtilis*.

3.2 Direct manipulation of guest BACs

As guest DNA replicates as part of the host genome, sequence modifications/conversions targeting regions in the *B. subtilis* genome are possible (Fig. 4). In addition to sequence conversions or small DNA insertions, the formation of systematic deletions was examined for a guest DNA derived from the mouse genome. The protocol to induce deletion shown in Fig. 5a applies a method that is the reverse of the method for integration; it uses two homologous recombinations. Two small DNA segments with deletion endpoints are connected so as to flank appropriate antibiotic resistance markers for *B. subtilis*. The region to be deleted was replaced by a marker gene via two homologous recombinations at the two flanking segments. Kaneko et al. (2003) succeeded in inducing four deletions ranging from 11 to 86 kbp in the mouse *jumonji* (*jmj*) gene locus (Fig. 5b). The preparation of DNA tools by using conventional gene engineering technologies in *E. coli* plasmids and their introduction into competent BGM are now routine. We do not think that such designed and well-controlled deletion formation is possible using standard BAC manipulation kits developed for *E. coli*. Other auxiliary tools, antibiotics resistance genes, and rare-cutting endonucleases, to apply and improve these protocols are presented in Chapter 4 below.

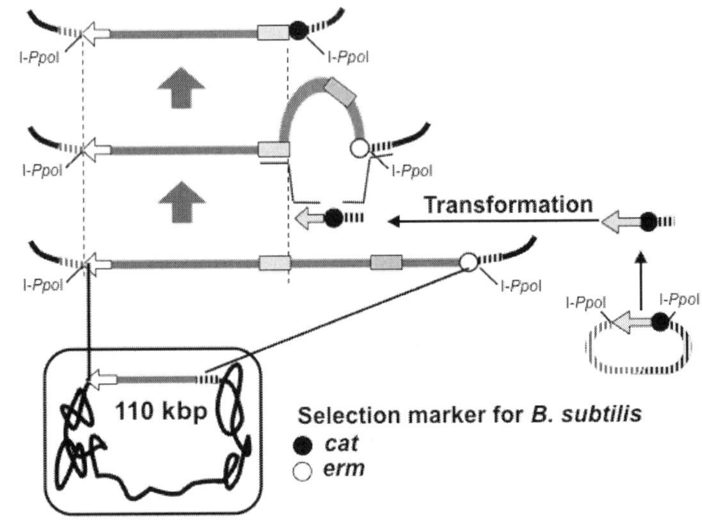

Fig. 5a. Concept underlying the formation of deletions.
Transformation using the optional small DNA fragments and antibiotic resistance markers for *B. subtilis* produces the designed deletion formation via homologous recombination.

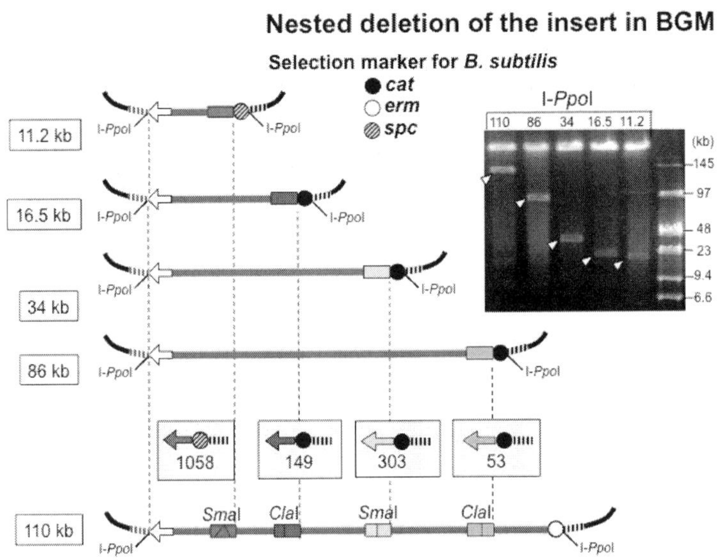

Fig. 5b. Systematic deletion formation from a mouse genomic region (110kb).
The BGM system makes it possible to conduct massive and systematic deletions. The picture on the right includes I-*Ppo*I fragments resolved by gel electrophoresis (open arrowheads with the size of the deletion indicated on the left). The BGM vector (4.2 Mb) migrates slowly.

3.3 Connection of two adjacent BACs in BGM

Itaya et al. (2008) documented that the integration of two partially overlapping fragments is possible in BGM. Overlaying the second on the first fragment in BGM resulted in elongation or connection. Each fragment is called a domino. If serial dominos are toppled, or all dominos are connected in BGM, reconstruction of the full-length guest DNA covered by these dominos completes. This concept was first realized by using dominos made in the pBR322-based plasmids, pCISP401(*cat*) and pCISP402(*erm*) in *E. coli*. Up to 31 dominos designed to completely cover the 135-kbp rice chloroplast genome produced a reconstructed full-length guest DNA (Itaya et al., 2008). The use of BACs instead of pBR322 should work in a similar manner. The scenario for connecting two adjacent BACs in a BGM vector is shown in Fig. 6a. However, selection markers present a problem in the preparation of BAC-dominos for immediate use. pBR-dominos are commonly prepared from PCR products of less than a few dozen kbp. However, BACs normally carry 100-kbp DNA fragments that exceed the limit of PCR-mediated amplification. Therefore at present we are forced to use BAC clones, such as commercially-available mouse genomic BAC libraries. However, their BAC vector does not possess *ab initio* selection markers for *B. subtilis*.

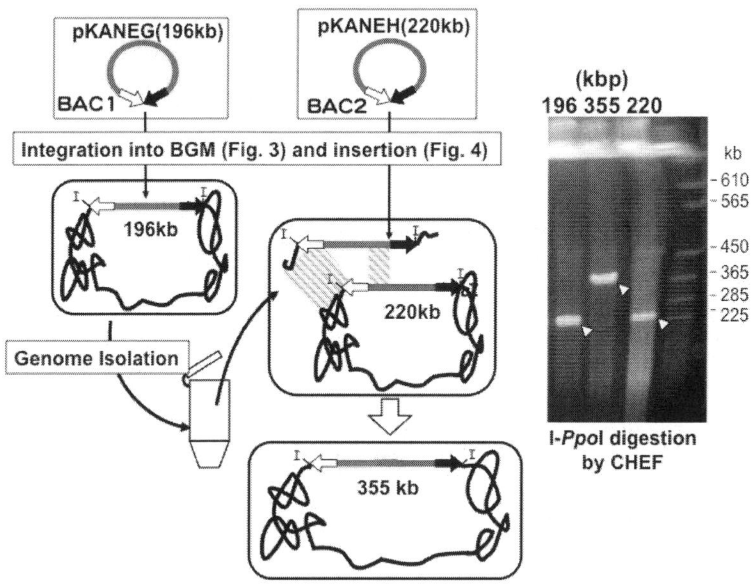

Fig. 6a. Connecting two adjacent BACs in BGM.
Two BAC clones, pKANEG (196 kb) and pKANEH (220 kb), cover the mouse genomic *jmj* region; there is a 60-kb overlap sequence. Each BAC was individually integrated into BGM. Transformation of the BGM carrying the 220-kb BAC2-DNA by using purified genomic DNA from another BGM carrying the 196-kb BAC1-DNA leads to homologous recombination between the 60-kb overlapping region and the sequence shared with the *B. subtilis* genome portion (see the splice boxes). This results in the production of a reconstructed mouse genomic *jmj* region (355kb). BAC insertion after I-*Ppo*I digestion was confirmed by agarose gel electrophoresis (right panel). The I-*Ppo*I recognition sequence is indicated by I.

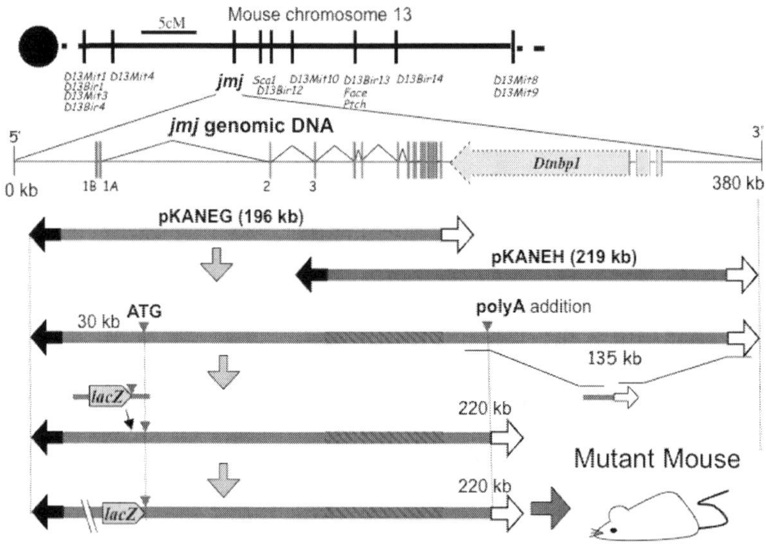

Fig. 6b. Summary of the current achievements made with the mouse a *jmj* gene.

As shown in Fig. 6b, the connection of two sequential guest BACs was first demonstrated for the two BACs covering the mouse gene *jumonji* locus (Kaneko et al., 2009). The two inserts, 196 kbp- and the 220 kbp guests of BGM, shared DNA approximately 60 kbp in size. While the sketch illustrating the connection looks simple, there are difficulties in using marker genes for *B. subtilis*. The combinatorial use of antibiotic markers, already reported by Kaneko et al. (2009) is omitted in Fig. 6a. Instead, the difference from the pBR322-based domino-connection is clearly shown: total genomic DNA isolated from one domino BGM was added to competent cells of the second domino BGM to force double homologous recombination. The transformants obtained by rational antibiotics selection produced a 355 kb-long connected BAC (Kaneko et al., 2009). A summary of our current achievements with the mouse *jmj* gene is presented in Fig. 6b; the most time- and labor-consuming step is the integration of unmarked BACs into the BAC-BGM. Besides these two elaborated experimental works using regular BACs, we have prepared BAC vectors that are designed to accomplish versatile aims in the BGM system. The two new BAC vectors, p108BGMC(*cat*) and p108BGME(*erm*), feature the *B. subtilis* markers shown in Fig. 3. Their presence should facilitate the domino-mediated elongation/reconstruction of BAC-based guest DNA (our unpublished data).

3.4 Implementation of sequence engineering in BACs (inversion)

What can we do with guest DNA? Figure 4 presents a list of possible modifications. Among them, techniques to induce the inversion of guest DNA appears as important as elongation. It is difficult to regulate the orientation of the inserted DNA in BACs. As this difficulty is frequently encountered at the construction of random BAC libraries, tools are needed to invert the insert present in BGM. The method and timing for the induction of large regions of the host *B. subtilis* genome have been described (Toda et al., 1996; Kuroki et al., 2007); they were applied to reverse the orientation of BAC inserts in BGM. Two partially overlapping fragments [ne] and [eo] derived from a neomycin resistant gene [*neo*] play an essential role; [ne] and [eo] are inserted at the terminus of the insert to be inverted in the BGM. Because these two fragments have an identical region designated [e], homologous recombination here produces two segments, [neo] and [e], that accompany the inversion of the intermediate insert between [ne] and [eo] (Toda et al., 1996). The inversion formation is always associated with the formation of [neo] and can be monitored by resistance to neomycin. This manipulation, theoretically simple but complex in its application, is one of the key technologies for BGM. We have already applied this tool to BACs covering other mouse genomic regions (unpublished findings).

3.5 Long-term storage of guest BACs in *B. subtilis* endospores

The BGM derived from *B. subtilis* is capable of forming spores reminiscent of plant seeds. Spores survive for a long time in unfavorable environments including aridity (Nicholson et al., 2000, 2002; Nicholson & Galeano, 2003; Takahashi et al., 1999; Benardini et al., 2003). *B. subtilis* spores readily start to germinate when moved back into nutrition-rich broth and start to grow immediately. We examined the stability of DNA inside spores (Kaneko et al., 2005) and found that the spores are stable for year at room temperature without requiring special devices. BGM spores may not be of value for use with small DNA that is synthesized and prepared routinely. However, as the size of the guest DNA increases, these spores should become more valuable. BGM spores should provide a low-cost, long-term reservoir for guest DNAs.

4. Heavily engineered BACs for downstream pipelines

Engineered DNA molecules must be delivered to other host systems for diverse applications. The use of BACs for purposes other than sequence determinations, for example, their use in mutagenesis studies on animal models was of limited success (Yang et al., 1997) because complex animal genes range from dozens to hundreds of kbp. They include factors, promoters, 5'untranslated regulatory sequences, introns, exons, splicing- and alternative splicing sites, 3'long terminal repeats, and polyA addition sites (Fig. 6b). Functional accessory sequences contain encoding small RNAs and non-coding RNA, and are regulated by, for example, methylation, histone-binding, and nucleosomes. The study of these complex mammal genes by reverse genetic approaches requires systems that facilitate the delivery of larger DNA than is used for microbial reverse genetics where gene units tend not to exceed kbp orders. For such studies, engineered BACs in BGM must be retrievable. A few methods for retrieving DNA from the *B. subtilis* genome are summarized in Fig. 7, three such methods have been applied to retrieve engineered BACs.

4.1 Fragment isolation using sequence-specific endonucleases

The simplest and most straightforward method involves (i) digestion by special endonucleases and (ii) subsequent isolation/purification of the engineered BAC. The use of two endonucleases facilitates the recognition of extremely infrequent sequences. They are: I-*Ppo*I for the 23-base sequence ATGACTCTCTTAA/GGTAGCCAAA, and I-*Sce*I for the 18-base sequence TAGGGATAA/CAGGGTAAT (Itaya, 2009). Indeed, two I-*Ppo*I recognition sequences are created in the *B. subtilis* genome to cut the BAC part out of the host genome (Figs. 5b and 6a). The linear DNA produced by I-*Ppo*I digestion is readily isolated from agarose gels resolved by pulsed-field gel electrophoresis. The handling of giant DNA longer than several hundred kbp has been difficult because of its fragility in test tubes. However, its use for microinjection into, for example, fertilized mouse eggs requires pure and undamaged DNA. We implemented and tested various steps to isolate undamaged giant DNA maintained in liquids. Our improved protocol that involves 2 gel electrophoresis runs and collection on a dialysis membrane made it possible to concentrate undamaged giant DNA. The successful purification of up to 220 kb mouse genomic DNA carrying reporter genes facilitated the creation of transgenic mice (manuscript in preparation). Because our method is technically simple it will be of great value in future studies.

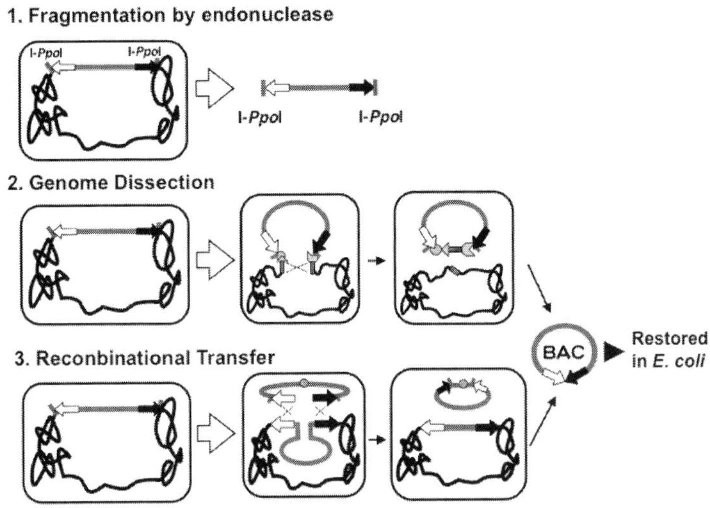

Fig. 7. Three methods to retrieve BAC-DNA from BGM.

1. Fragmentation using endonucleases is simple. Improvements for the isolation step are described in the text.
2. Only one report describing genome dissection has been published (Itaya & Tanaka, 1997). The potential for isolating DNA above 300-kbp is described.
3. For recombinational transfer see the text and refer to Tsuge & Itaya (2000) and Itaya & Tsuge (2011). Although the process appears to be complicated, the BAC insert can be copied in the BGM and pasted into a linear plasmid to complete circularization. This process yields a plasmid carrying the copied DNA segment.

4.2 Genetic dissection of the guest portion from the *B. subtilis* genome

The second method, genome dissection, largely depends on *B. subtilis* genetic systems. Although this method appears complex at first glance because of the molecular apparatus briefly described in Fig. 7, it is in fact simple. Intrachromosomal homologous recombination between the two DNA repeats makes it possible to physically disconnect the BAC segments. As the disconnected DNA was designed to carry a DNA replication origin site (*oriN*) different from *oriC* of the chromosome, it starts replicating autonomously as a plasmid independent from the chromosome. Itaya and Tanaka (1997) reported a 300-kbp DNA fragment that manifested superb genetic stability. Despite the potential to produce circular

DNA larger than 300 kbp by this method, its application has been restricted by the rather complicated procedures involved in its creation (Itaya, M., unpublished observations).

4.3 Retrieval by copying segments of the *B. subtilis* genome

The third method involves a yet more complicated genetic process referred to as *Bacillus* Recombinational Transfer (BReT) (Tsuge & Itaya, 2001; Kaneko et al., 2005). Indeed, BReT relies on homologous recombination in *B. subtilis* and should be as simple as the genetic disconnection procedure described above. Different from the disconnection procedure, the engineered BAC parts from the genome are copied and pasted into the existing plasmid. DNA retrieval from the genome to the plasmid involves a reverse direction of the DNA delivery into a host pBR322 or BAC as illustrated in Fig. 1b. The complete circular form yielded by the BReT pathway, selection with plasmid-linked markers followed by standard extraction of plasmid DNA, resulted in the purification of complete recombinant genomes of lambda (Tsuge & Itaya, 2001), organelle genomes from mitochondria and chloroplast (Itaya et al., 2008; Yonemura et al., 2007), and several BACs (Kaneko et al., 2005). The engineering of organelle genomes, mitochondria, and chloroplasts opens an exciting field because typical organelle genome sizes, ranging from 100-200 kbp for chloroplast, are below the carrying capability of BACs. Although there are currently no reliable technical tools to deliver them back to natural cells, the availability of circular freely engineered mitochondria or chloroplast genomes raises the need for such tools (Gibson et al., 2008, 2010; Itaya et al., 2008; Itaya, 2010).

5. Choice of the two hosts for BAC engineering

The use of BACs addresses different aims and goals in different cellular systems and some disadvantages must still be overcome. One seeming disadvantages is that our *B. subtilis* system is still too new for wide dissemination. Examples of BACs that have been manipulated in BGM systems to date are listed in Table 1. Advantages are that the *B. subtilis* host system is obviously superior to BAC technologies that make use of the *E. coli* system (Fig. 2). We focused on using BACs in our mouse genetics studies; investigations of gene loci other than *jmj* (Fig. 6b) are in progress. Another advantage is that a total of 9 antibiotic resistance gene cassettes are currently available for *B. subtilis* selection: neomycin (*neo*), spectinomycin (*spc*), chloramphenicol (*cat*), tetracycline (*tet*), erythromycin (*erm*), blasticidin S (*bsr*), kanamycin (*kam*), phleomycin (*phl*), and hygromycin (*hyg*). These antibiotic resistance genes will greatly facilitate the development of multipurpose BGM vectors that allow all desired manipulations. Although examples for the practical application of BAC technologies remain limited, we are in the process of producing a BAC-BGM kit that involves a simple protocol. The next generation of BGM should be accompanied by handy manuals so that even non-expert users of *B. subtilis* can perform routine experiments based on a basic understanding of the principles that are the foundation of BGM. The rapid preparation and assembly of fragments in the host is key for simplifying necessary procedures and for shortening the time required for the production of engineered fragments. From this perspective, we think that the BGM system is an invaluable tool for the creation of accurately designed guest DNA that does not rely on existing BAC modification kits in *E. coli*. However, the role of *E. coli* as an initial producer of the BACs prior to BGM will remain unaltered. Initially, our new BAC vectors, p108BGMC(*cat*) and p108BGME(*erm*), can be expected to replace the currently widely-used BAC vectors that lack markers for BGM.

DNA source from	insert*2 size (kbp)	Accommodation in*3 BGM by Fig.1b	Retrieval in plasmid as Fig.7	Restored in*4 *E. coli*
***Arabidopsis thaliana* mtDNA**†1				
F1O22	101	**Yes**	**Yes**	**Yes**
F2L14	100	**Yes**	**Yes**	**Yes**
F3A21	100	**Yes**	**Yes**	**Yes**
F3B3	115	**Yes**	**Yes**	**Yes**
F4C19	100	**Yes**	**Yes**	**Yes**
F4I8	100	**Yes**	**Yes**	**Yes**
F4O20	144	**Yes**	**Yes**	**Yes**
F6A21	115	**Yes**	**Yes**	**Yes**
F6A7	105	**Yes**	**Yes**	**Yes**
F7I2	85	**Yes**	**Yes**	**Yes**
F7J2	98.5	**Yes**	**Yes**	**Yes**
F9J20	120	**Yes**	**Yes**	**Yes**
F10J8	90	**Yes**	**Yes**	**Yes**
F10L13	89.5	**Yes**	**Yes**	**Yes**
F10F17	100	**Yes**	**Yes**	**Yes**
F11E12	80	**Yes**	**Yes**	**Yes**
F13E8	90	**Yes**	**Yes**	**Yes**
F13O24	115	**Yes**	**Yes**	**Yes**
Mouse *Jmj* gene (Chromosome 13)				
pKANED†2	150	**Yes**	(NT)	(NT)
	(Deletion of pKANED as in Fig.5a and 5b)			
	110	**Yes**	(NT)	(NT)
	86	**Yes**	(NT)	(NT)
	34	**Yes**	**Yes**	**Yes**
	16.5	**Yes**	**Yes**	(NT)
	11.2	**Yes**	**Yes**	**Yes**
pKANEG†3	198	**Yes**	(NT)	(NT)
pKANEH†3	219	**Yes**	(NT)	(NT)
(Connection: pKANEG +pKANEH in Fig.6a)				
	355	**Yes**	(NT)	(NT)
Mouse *TS* gene DNA (Chromosome 11) †4				
p185A21	120	**Yes**	(NT)	(NT)

*1BAC clones in literature *2Estimated by gel electrophoresis. *3Confirmed by Southern blot analysis using original BAC clone as probes. *4Used for transformation of *E. coli* strains DH1 or DH10B. Refereed by †1 Kaneko et al., 2005, †2 Kaneko et al., 2003 †3 Kaneko et al., 2008, and †4 Itaya et al., 2000.

Table 1. BAC clones*1 handled by BGM systems

6. Future perspectives

Based on available data, we think that the use of *B. subtilis* will facilitate the rapid and efficient engineering of BACs. While both *E. coli* and *B. subtilis* are excellent BAC hosts, several experimental steps require that DNA be manipulated outside these hosts. Steps involved in the delivery of DNA from *E. coli* to BGM and in the isolation of engineered guest DNA render the fragments maintained in liquid vulnerable to damage due to the intrinsic physicochemical nature of these liquids.

Elsewhere we reported that the conventional delivery step may be replaced (Kaneko & Itaya, 2010a, 2010b; Itaya & Kaneko, 2010). We fortuitously found that plasmid DNA remains intact for a while after the induction of lambda phage of *E. coli*. While *E. coli* was lysed and its genome degraded, the co-existing plasmid remained intact and available for transfer to competent *B. subtilis*. The stability of plasmids in the lambda-induced *E. coli* lysate also applied to BACs (Kaneko & Itaya, 2010a, 2010b; Itaya & Kaneko, 2010). Consequently, it is no longer necessary to attempt to prepare undamaged BACs in test tubes.

The techniques involved in the isolation of undamaged DNA from electrophoresis gels have been improved and applied to fragments of DNA measuring up to 220 kbp. In the post-genome-sequencing era, reverse genetics using designed and manipulated BACs and BGM will play an essential role in the construction of various genetic mutants.

7. Concluding remarks

Genome-sequence determination techniques are now applied to a vast range of species. In contrast, techniques to engineer genomes, even of bacteria far smaller than those of higher eukaryotes, remain scarce. Recently, two independent laboratories, one of them ours, demonstrated that whole bacterial genomes can be cloned/manipulated (Itaya et al., 2005). This technological breakthrough has dramatically changed various aspects of genome engineering. However, these cutting-edge technologies are not yet widely applied because they remain labor-intensive with respect to their use in genomes and because the cost of producing correct genomes is high. Besides introducing the reader to the cloning of whole bacterial genomes, this chapter aimed at describing multipurpose systems by which DNA can be routinely engineered. Giant guest DNA cloned in the BGM vector will play a significant role in versatile gene/genome delivery systems. The goal of genome engineering is not only the propagation of microbes but also the engineering of cells addressed by different branches of the life sciences. BAC-based cloning in *E. coli* (Shizuya et al., 1992), one of many DNA cloning technologies, has been revisited in efforts to develop essential cutting-edge tools that can be applied in the BAC-BGM system. The first BAC transfer to- and manipulation in the BGM system was achieved as little as 10 years earlier (Itaya et al., 2000). Subsequent publications demonstrated that once the BAC guest was incorporated into the BGM host, modified guest DNA and the novel homologous recombination activity of *B. subtilis* made this possible due to the amazing structural stability of guest DNA. Furthermore, depending on experimental requirements, the structure of recovered fragments can be circular or linear. Emerging genome synthesis technologies will yield giant DNA equivalent to the size of bacterial genomes (Gibson et al., 2008, 2010; Gibson, 2011). We are entering a new era of novel synthetic biology that relies on the synthesis/construction of designed biomaterials. With the aid of newly designed genes and proteins, *de novo* synthetic pathways can be expected to produce yet unknown substances and with the aid of giant DNA, many genes can be delivered simultaneously.

While genome synthesis is gradually changing the way of thinking of researchers in the life sciences, the introduction, maintenance, and manipulation of DNA fragments in BACs will continue to shed light on other host- and biological systems. The *de novo* synthesis of sufficiently large DNA in BACs is only a matter of time and it remains to be seen whether current gene engineering systems will continue to be relevant. We will continue to dedicate our careers to working with DNA pieces irrespective of their size and to shedding light on their divergent applicability to biological systems used in research.

8. Acknowledgement

We thank Dr. Hirota for the unpublished studies that used BGM-based BAC clones in the production of murine genetic mutants.

9. References

Ara, K.; Ozaki, K.; Nakamura, K.; Yamane, K.; Sekiguchi, J. & Ogasawara, N. (2007) *Bacillus* minimum genome factory: Effective utilization of microbial genome information. *Biotechnol Appl Biochem.* Vol. 46, pp.169-178.

Benardini, JN.; Sawyer, J.; Venkateswaran, K. & Nicholson, WL. (2003) Spore UV and acceleration resistance of endolithic *Bacillus pumilus* and *Bacillus subtilis* isolates obtained from Sonoran desert basalt: Implications for lithopanspermia. *Astrobiology.* Vol.3, pp.709-717.

Frengen, E.; Weichenhan, D.; Zhao, B.; Osoegawa, K.; van Geel, M. & de Jong, PJ. (1999) A modular, positive selection bacterial artificial chromosome vector with multiple cloning sites. *Genomics* Vol.58, pp.250-253.

Gibson, DG. (2011) Enzymatic assembly of overlapping DNA fragments. *Method Enzymol.* Vol. 498, pp.349-361.

Gibson, DG.; Benders, GA.; Andrews-Pfannkoch, C.; Denisova, EA.; Baden-Tillson, H.; Zaveri, J.; Stockwell, TB.; Brownley, A.; Thomas, DW.; Algire, MA.; Merryman, C.; Young, L.; Noskov, VN.; Glass, JI.; Venter, JC. ; Hutchison, CA 3rd. & Smith HO. (2008) Complete chemical synthesis, assembly, and cloning of a *Mycoplasma genitalium* genome. *Science* Vol.319, pp.1215–1220.

Gibson, DG.; Glass, JI.; Lartigue, C.; Noskov, VN.; Chuang, RY.; Algire, MA.; Benders, GA.; Montague, MG.; Ma, L.; Moodie, MM.; Merryman, C.; Vashee, S.; Krishnakumar, R.; Assad-Garcia, N.; Andrews-Pfannkoch, C.; Denisova, EA.; Young, L.; Qi, ZQ.; Segall-Shapiro, TH.; Calvey, CH.; Parmar, PP.; Hutchison, CA 3rd.; Smith, HO. & Venter, JC. (2010) Creation of a bacterial cell controlled by a chemically synthesized genome. *Science*, Vol. 329 pp.52-56.

Hanahan, D. (1983) Studies on transformation of *Escherichia coli* with plasmids. *J. Mol. Biol.* Vol.166, pp.557-580.

Hardy, S.; Legagneux, V.; Audic, Y. & Paillard, L. (2010) Reverse genetics in eukaryotes. *Biol Cell.* Vol.102 pp.561-580.

Itaya, M. (1993) Integration of repeated sequences (pBR322) in the *Bacillus subtilis* 168 chromosome without affecting the genome structure. *Mol Gen Genet.* Vol.241, pp.287-297.

Itaya, M. (2009) Recombinant genomes, In: *Systems Biology and Synthetic Biology,* Fu, P. & Panke, S. (Eds.), pp.155-194, John Wiley & Brothers, Inc. New Jersey.

Itaya, M. (2010) A synthetic DNA transplant. *Nat. Biothechnol,* Vol.28, pp.687-689.

Itaya, M.; Fujita, K.; Kuroki, A. & Tsuge, K. (2008) Bottom-up genome assembly using the *Bacillus subtilis* genome vector. *Nat. Methods,* Vol. 5, pp.41-43.

Itaya, M. & Kaneko, S. (2010) Integration of stable extra-cellular DNA released from *Escherichia coli* into the *Bacillus subtilis* genome vector by culture mix method. *Nucleic Acids Res.* Vol.38, pp.2551–2557.

Itaya, M.; Shiroishi, T.; Nagata, T.; Fujita, K. & Tsuge, K. (2000) Efficient cloning and engineering of giant DNAs in a novel *Bacillus subtilis* genome vector. *J. Biochem.* Vol.128, pp.869 - 875.

Itaya, M. & Tanaka, T. (1991) Complete physical map of the *Bacillus subtilis* 168 chromosome constructed by a gene-directed mutagenesis method. *J. Mol. Biol.* Vol.220, pp.631-648.

Itaya, M. & Tanaka. T. (1997) Experimental surgery to create subgenomes of *Bacillus subtilis* 168. *Proc. Natl. Acad. Sci., USA.* Vol.94, pp.5378-5382.

Itaya, M. & Tsuge, K. (2011) Construction and manipulation of giant DNA by a genome vector. *Methods Enzymol.* Vol.498, pp.427-447.

Itaya, M.; Tsuge,K.; Koizumi, M. & Fujita, K. (2005) Combining two genomes in one cell: Stable cloning of the *Synechocystis* PCC6803 genome in the *Bacillus subtilis* 168 genome. *Proc. Natl. Acad. Sci., USA.* Vol.102, pp.15971-15976.

Kaneko, S.; Akioka, M.; Tsuge, k. & Itaya, M. (2005) DNA shuttling between plasmid vectors and a genome vector: Systematic conversion and preservation of DNA libraries using the *Bacillus subtilis* genome (BGM) vector. *J. Mol. Biol.* Vol.349 pp.1036-1044.

Kaneko, S. & Itaya, M. (2010a) Designed horizontal transfer of stable giant DNA released from *Escherichia coli. J. Biochemistry* Vol.147, pp.819-822.

Kaneko, S. & Itaya, M. (2010b) Stable extracellular DNA: a novel substrate for genetic engineering that mimics horizontal gene transfer in nature In: *Extracellular Nucleic Acids. Nucleic Acids and Molecular Biology Series 25,* Kikuchi, Y. & Rykova, E.Y. (Eds), pp.39-53, Springer.

Kaneko, S.; Takeuchi, T., & Itaya, M. (2009) Genetic connection of two contiguous bacterial artificial chromosomes using homologous recombination in *Bacillus subtilis* genome vector. *Journal of Biotechnology,* Vol.139, pp.211-213.

Kaneko, S.; Tsuge, K.; Takeuchi, T. & Itaya, M. (2003) Conversion of submegasized DNA to desired structures using a novel *Bacillus subtilis* genome vector. *Nucleic Acids Res.* Vol.31, e112.

Kidane, D. & Graumann, PL. (2005) Intracellular protein and DNA dynamics in competent Bacillus subtilis cells. *Cell.* Vol.122. pp.73-84.

Kovács AT, Smits WK, Mirończuk AM, Kuipers OP. (2009) Ubiquitous late competence genes in Bacillus species indicate the presence of functional DNA uptake machineries. *Environ Microbiol.* Vol.11. pp.1911-1922.

Kuroki, A.; Ohtani, N.; Tsuge, K.; Tomita, M. & Itaya, M. (2007) Conjugational transfer system to shuttle giant DNA cloned by the *Bacillus subtilis* genome (BGM) vector. *Gene* Vol.399, pp.72-80.

Mandel, M. & Higa, A. (1970) Calcium-dependent bacteriophage DNA infection. *J. Mol. Biol.* Vol.53 pp.159-1624.

Nicholson, WL.; Fajardo-Cavazos, P.; Rebeil, R.; Slieman, TA.; Riesenman, PJ.; Law, JF. & Xue, Y. (2002) Bacterial endospores and their significance in stress resistance. *Antonie Van Leeuwenhoek.*Vol.81, pp.27-32.

Nicholson, WL. & Galeano, B. (2003) UV resistance of *Bacillus anthracis* spores revisited: validation of *Bacillus subtilis* spores as UV surrogates for spores of *B. anthracis* Sterne. *Appl Environ Microbiol.* Vol.69, pp.1327-1330.

Nicholson, WL.; Munakata, N.; Horneck, G.; Melosh, HJ. & Setlow, P. (2000) Resistance of *Bacillus* endospores to extreme terrestrial and extraterrestrial environments. *Microbiol Mol Biol Rev.* Vol.64, pp.548-572.

Osoegawa, K.; Mammoser, AG.; Wu, C.; Frengen, E.; Zeng, C.; Catanese, JJ. & de Jong, PJ. (2001) A bacterial artificial chromosome library for sequencing the complete human genome. *Genome Res.* Vol. 11, pp.483-496.

Shizuya, H.; Birren, B.; Kim, U. J.; Mancino, V.; Slepak, T.; Tachiiri, Y. & Simon, M. (1992) Cloning and stable maintenance of 300-kilobase-pair fragments of human DNA in *Escherichia coli* using an F- factor-based vector., *Proc. Natl. Acad. Sci. USA* Vol.89, pp.8794-8797.

Spizizen, J. (1958) Transformation of biochemically deficient strains of *Bacillus subtilis* by deoxyribonuclease. *Proc. Natl. Acad. Sci. USA* Vol.44, pp.1072-1078.

Takahashi, N.; Hieda, K.; Morohoshi, F. & Munakata, N. (1999) Base substitution spectra of nalidixylate resistant mutations induced by monochromatic soft X and 60Co gamma-rays in *Bacillus subtilis* spores. *J Radiat Res (Tokyo).* Vol.40, pp.115-124.

Toda, T.; Tanaka, T. & Itaya, M. (1996) A method to invert DNA segments of the *Bacillus subtilis* 168 genome by recombination between two homologous sequences. *Biosci Biotechnol Biochem.* Vol.60, pp.773-778.

Tsuge, K. & Itaya, M. (2001) Recombinational transfer of 100-kb genomic DNA to plasmid in *Bacillus subtilis* 168. *J. Bacteriol.* Vol.183, pp.5453-5456.

Yang, XW.; Model, P. & Heintz, N. (1997) Homologous recombination based modification in *Escherichia coli* and germline transmission in transgenic mice of a bacterial artificial chromosome. *Nat Biotechnol.* Vol.15 pp.859-865.

Yonemura, I.; Nakada, K.; Sato, A.; Hayashi, J.; Fujita, K.; Kaneko, S. & Itaya, M. (2007) Direct cloning of full-length mouse mitochondrial DNA using a *Bacillus subtilis* genome vector. *Gene,* Vol.391, pp.171-177.